D0621094

622.18282
H879i

cards ordered

→ Interval Velocities from Seismic Reflection Time Measurements

by

Peter Hubral
Bundesanstalt für Geowissenschaften
und Rohstoffe
Hannover, West Germany

and

Theodor Krey
Consultant
Prakla-Seismos
Hannover, West Germany

NDD
TN269

OCLC
6479371

Edited by

Kenneth L. Larner
Western Geophysical Company
Houston, Texas

54222

ISBN 0-931830-13-3

Library of Congress Catalog Card Number: 80-50915

Society of Exploration Geophysicists
P.O. Box 3098, Tulsa, Oklahoma 74101

©1980 by the Society of Exploration Geophysicists
All rights reserved

Published 1980

Printed in the United States of America

CONTENTS

Editor's Preface

Over the years, ray theory has furnished the exploration geophysicist with most of the working tools for understanding and interpreting events observed on reflection seismic sections. Even today, notwithstanding the pace at which the more powerful acoustic wave theory is introducing its new tools, ray theory, in the hands of Peter Hubral and Theodore Krey, retains its preeminence for providing insights into fundamental problems in reflection seismology. Professor Krey's earlier contributions are part of ray theory's rich heritage. Alongside C. Hewitt Dix and Hans Durbaum, he elucidated relationships between interval velocity and observed reflection moveout.

In the 1950s, these relationships were developed for comparatively simple subsurface models consisting either of horizontal layers or of layers bounded by plane dipping interfaces, all having the same strike. In this monograph, ray theory resurfaces to tackle problems involving models of considerably more complexity. In these models, layer interfaces are curved and oriented arbitrarily in three-dimensional space. Moreover, the problems treated here are at the core of issues of primary concern in current processing and interpretation. Prominent among them are (1) the migration of data from complex media, i.e., the depth migration problem, (2) the determination of velocities for performing migration in two and three dimensions, and (3) the estimation of interval velocity from surface seismic measurements, again for three-dimensionally complex media. The reader will also find an analysis of the field survey configurations that provide the necessary and sufficient moveout information for determining stacking velocities and migration velocities and, hence, for obtaining subsurface velocities in three dimensions.

The approaches throughout the monograph are analytic. Although results for the simpler models are expressed in closed form, the authors in general have opted for a recursive form of solution that is eloquent in its generality and in the physical insights it provides. Thus, for example, the tricky inverse problem of estimating interval velocity from observations of traveltime reduces to a backward (and downward) propagation of wavefronts to their subsurface sources. Within this recursive framework, the intimate interrelationships among stacking, migration, velocity estimation, wavefront curvature, and focusing and defocusing of energy are clearly revealed.

This monograph could be subtitled "Essay on the Art of the Second-Order Approximation as Applied to Traveltime Measurements." Throughout, the analytic approaches derive their tractability from adherence to the practical. Offsets, whether between source and receiver or between image ray location and neighboring positions, must be small relative to the distances traveled along reflected raypaths. The price for accomplishing so much with the analytic approach is that subsurface models, for all their relative complexity, must retain a degree of simplicity; interfaces must be locally smooth; strictly, faults and unconformities are excluded. Such limitations are not completely satisfactory for estimating subsurface velocity structure in the detail required for lithologic determination and for accurate depth migration. On the other hand, any alternative numerical approach capable of treating a complex subsurface properly will at best be unable to provide the unity, perspective, and insight contained in the analytic approach of this work.

The dominant messages of this monograph are that stacking velocity and migration velocity need not be the same; that stacking velocity is not indentical to root-mean-square velocity; and that where geologic structure is complex, the venerable Dix equation necessarily yields unacceptable values of computed interval velocity. The reader will feel rewarded, however, to find that stacking velocity and migration velocity are blood relatives, that stacking velocity often approximates a generalized form of root-mean-square velocity, and that $T^2 - X^2$ analysis can successfully remain the basis for interval velocity determination when geologic structure is complex.

The Editor acknowledges his appreciation to the members of the Society of Exploration Geophysicists' Publication Committee, especially to Manus Foster who as Editor of GEOPHYSICS saw the wisdom of combining the many related works of Drs. Hubral and Krey, both those published in GEOPHYSICS and those submitted for publication, into a single, unified treatment. Special thanks also go to Belynda Bland, Editorial Assistant on the Publication Staff of the Society, who helped in proofreading and editing for style as well as in carrying the manuscript through the stages required for publication. The editor also acknowledges the constant support of Jerry Henry, Publication Manager for the Society.

Kenneth L. Larner

Houston, Texas
April 1, 1980

Preface

Various types of velocity have been introduced during the recent history of exploration seismology. They have become increasingly important for interpretation, evaluation, and processing of seismic reflection data used in exploring for hydrocarbons and other natural resources. Prominent among these quantities are stacking velocity, migration velocity, normal moveout velocity, and root mean square (RMS) velocity. Each of these types is derived from seismic reflection measurements—often with a high degree of accuracy. In different ways, they all involve the velocity of local seismic wave propagation (typically *P*-wave velocities); this is a parameter of great interest to explorationists and seismic interpreters. After initial successes with application of the process of "time migration" to seismic reflection measurements, it is becoming increasingly obvious that further essential improvements of the method (particularly with respect to migrating reflections properly into depth) will depend largely on an exact knowledge of true local velocities.

No other geophysical parameter is a more integral factor in the reflection seismic process than is local velocity. Its spatial distribution accounts for all distortion of wavefronts. It determines how, where, and when wavefronts propagate. Together with bulk densities and with absorptive and dispersive medium properties, it influences the relative amplitudes of recorded motion.

Various methods of velocity determination are currently employed in reflection seismology. Our purpose is to clarify the theory and limitations of those methods that are based on the analysis of recorded traveltime measurements for both reflected and diffracted seismic arrivals. This monograph is, therefore, written largely for exploration seismologists who desire a general appreciation of problems related to computing interval velocities and to solving the inverse traveltime problem in reflection seismology. Some of our material constitutes a review of a great many publications on this subject.

As yet undiscovered natural resources are confined increasingly to the more complex and more subtle geologic environments conducive to the trapping of deposits. Confronted with the task of deriving fundamental parameters (of use in locating these resources) from large amounts of processed seismic reflection data, exploration seismologists are now assisted largely by digital computers and "black box" programs which often involve sophisticated algorithms. They are thus able to concentrate on more detailed interpretation and expend more effort on the edge of the unknown—where experience and imagination are required to infer other than the obvious and where seismic data must be integrated with other corroborative geophysical and geologic information.

Geophysicists are well aware of the importance of seismic velocities to processing and interpretation. To achieve meaningful results, they should be familiar with the fundamental principles that underly the processing—in particular, processing to derive velocity information.

Any seismic data processing performed in a computer is based on some mathematical model. Specifically, accurate velocity determination requires not only quality traveltime measurements but also choice of appropriate models for the actual geology and for the seismic process. Even novices soon realize this. In general, the accuracy required for models and measurements depends on the particular use to which the velocity information is put. Velocities play essential roles in various aspects of seismic processing [e.g., common-datum-point (CDP) stacking, time migration, identification of multiples, and compensation of amplitudes for the effect of geometric spreading]. Apart from their importance in processes such as time-to-depth conversion, modeling, and migration, velocities themselves can be of great lithological value; while aiming for definition of both vertical and lateral velocity variations, one may expose stratigraphic subtleties related to, for instance, hydrocarbon presence, overpressured zones, sand-shale ratios, density, and porosity.

As no all-embracing model can exist for velocity determination, one must always find compromises between computational simplicity and physical reality. Explorationists aware of such compromises and the inherent model assumptions (necessarily leaving open a gap between theory and practice) can better appreciate the degree of confidence they can assign to seismic velocities. They can avoid being led astray by venturing beyond the limits of the theory; and they will refrain from various unjustified transgressions of the theory which are often forced upon them in the name of automation.

When seismic signals from a near-surface explosion return from their travels through the earth, the recorded reflection seismogram presents a highly mixed conglomerate of events. The desired information about subsurface velocities is embedded in the curvatures of diverging wavefronts and observed traveltime functions. Analysis of curvatures of traveltimes—in particular the normal moveout (NMO) of a CDP gather—has long been (Green, 1938) and still is the most important starting point for computing interval velocities. The methods described here are largely based on such analysis.

The intricate relationship between primary reflection amplitude and acoustic impedances can also be exploited for the estimation of seismic velocities. The computation of so-called *pseudosonic logs*, for example, makes use of this relationship. The pseudosonic log appears to be an increasingly more important tool for refining interval velocities obtained from NMO and diffraction arrivals. The theory upon which the computation of pseudosonic logs is based is not discussed here. Its recent success, however, has made us aware of the many facets encompassed by seismic velocities. Also, no mention will be made of the many traveltime inversion methods developed in refraction seismology. Such methods are often plagued with a basic nonuniqueness of the inversion problem which does not exist for the direct inversion methods described here.

In most geophysical interpretation theories, one has to resort to some type of iterative searching technique. One supplies a first guess, and, by linearizing the problem, finds corrections for the parameters that are to be established. It should be stressed that not many geophysical inversion problems lend themselves to a noniterative unique solution as is the case for the methods of computing interval velocities that we present here.

The primary topics of this monograph are the analyses of the properties and uses of stacking and migration velocities. Using ray-theoretical concepts only, we show how these velocities relate to structure and lithology and how they lend themselves to the computation of interval velocities. We will demonstrate their significance in time migration and time-to-depth migration processes. We will indicate how spread length, profile azimuth, and statics can affect the quality of derived velocity estimates. Particular emphasis is put on extending the Dix algorithm to complex, three-dimensional (3-D) layered models and to applying the resulting more general algorithm to the computation of interval velocities from stacking and migration velocities. Convinced that the geology is not part of a one-dimensional (1-D) or a two-dimensional (2-D) world, we discuss, whenever possible, problems in 3-D space.

The authors are greatly indebted to Prof. V. Cěrveny of the Charles University (Prague) and to Dr. I. Psenčik of the Geophysical Institute of the Czechoslovakian Academy of Science (Prague) for reading the manuscript and providing valuable comments.

1 Introduction

A cursory look at a continuous velocity log (CVL) provides sufficient evidence that hydrocarbons and other natural resources are certainly not deposited in simple geologic environments consisting of only a few homogeneous sedimentary layers bounded by first-order interfaces. Rather, resources typically are found within highly stratified layers which can be folded or faulted and can wedge out in various directions or be in unconformable contact with other subsurface masses. In addition, the various properties affecting seismic wave propagation within the earth may often change laterally within layers.

The true geology in all details can never be completely recovered from seismic reflections. Rather, one ought to aim for the recovery of a suitable *subsurface model* that provides all desired information related to expected resources and is reasonably consistent with other available surface or borehole measurements. In the past, models were greatly simplified. In the future they will become more complex, and one can expect them to provide a more equitable balance between *theory* and *observation*.

Deriving an accurate subsurface model from surface measurements, i.e., solving the *inverse problem*, is certainly no simple task. The process is particularly difficult where the geology (and, hence, the seismic trace) is complicated. Therefore, it is not surprising that many complementary methods have evolved which all aim for inverse solutions: estimates of *images* of the subsurface.

Various seismic waves are generated and recorded in response to a near-surface explosion. Some arise in close sequences of layers (short peg-leg multiples) or are generated at faults, discontinuities, strong interface curvature, and inhomogeneities (diffractions). Others occur in existing water layers (reverberations) or result from waves obliquely incident upon reflecting and refracting interfaces (mode-converted waves). A seismic trace is generally looked upon as a sequence of overlapping wavelets that result from a variety of paths. These wavelets are recorded at selected positions on or near the earth's surface. The recorded traces provide desired and redundant information about the subsurface. That information is concealed in *traveltimes* and *amplitudes.*

Seismic wave theory is concerned with the study of the relationship between subsurface parameters and expanding wave fields. It relies largely on solving the elastic (or scalar) wave equation for more or less complex models by considering certain initial, boundary, and interface conditions. The theory that deals with finding solutions of the wave equation for very high-frequency signals is called *geometrical optics*. Most wave propagation concepts used here belong to this theory. However, only those concepts that are considered to be of help in solving the inverse seismic problem directly (to the extent that this can be achieved with traveltime considerations alone) are treated in this monograph.

Though the tools of geometrical optics form the foundation for the computation of interval velocities and many other interpretation processes, their use is not always valid. In fact, a successful application of rays to seismic problems must be based on certain rules that are well laid down in the asymptotic ray method (see 4.5.3). Rays, for example, fail to describe properly both diffraction phenomena and interference of low-frequency waves. Moreover, they also generally fail to give an accurate account of amplitude or wave motion.

Because the process of *time migration* is based largely on the diffraction of seismic waves, ray theory has long been neglected as an interpretational aid for time-migrated reflections. As we shall see, however, the *normal ray*, descriptive of unmigrated stacked data, has a counterpart, the *image ray*, for describing time-migrated data. As a consequence, we use the image ray as the key for computing interval velocities from migration velocities. Migration velocities relate to image rays much as stacking velocities relate to normal rays. Rays, reflection points, point sources, and point scatters are nonexistent in actual wave propagation. The fact that these concepts are so basic in seismic exploration reflects partly the *dualism* that exists between ray and wave theory. Energy always returns to a seismometer from a large area [on the order of ten square wavelengths (Woods, 1975)] on an interface. It is never induced or returned from a point object, nor is it transported within an infinitely thin ray tube that encompasses a central ray.

Ray tracing frequently requires solving systems of ordinary differential equations. Readers not too familiar with this particular aspect of mathematics will be pleased to find that the theory used in this work does not call upon solving differential equations. We avoid differential equations and arrive at exact analytical inverse traveltime solutions because, throughout most of the monograph, we model the earth by a system of constant-velocity layers. We justify this restriction on grounds that (1) reflection data provide insufficient information for unique determination of velocity variations within layers, and (2) the models that we treat here represent a useful and instructive generalization relative to the more restrictive models treated previously.

The seismic ray method is not the only discipline that contributes to solving the inverse traveltime problem. Other aspects of exploration seismology are equally important. In particular, the high precision with which seismic signals can be recorded and processed nowadays, as well as the new standards of detailed care that are used in their analysis (Anstey, 1977) are a necessary prerequisite to the successful implementation of the methods described here.

Stacked or time-migrated sections are "processed seismic products" which are intermediate to the original seismic traces and the desired cross-section. They usually provide only an *approximate image* of the earth. They are, at most, deconvolved in the very general sense that unwanted information (e.g., multiples, diffractions) has been suppressed and desired information (e.g., primaries) enhanced. Seismic interpreters are thus better able to tell which energy is due to a *primary reflection* and which is not. Stacked and time-migrated sections still do not tell an interpreter all he desires to know. It remains to infer from them an exact image of the earth—a task often left to seismic interpreters.

There will most certainly come a time when seismic interpreters are no longer content with predominantly well-stacked or time-migrated sections. They will undoubtedly ask more and more for—what they really want—sections that represent a more complete cross-section through the earth and which are superimposed by a maximum amount of easily understandable extracted subsurface parameters. Since traveltimes are affected only by velocities, they, in turn, can be used only to recover subsurface parameters which influence the kinematic behavior of seismic waves. These parameters include subsurface velocity and depth, dip, strike, and curvature of reflecting horizons. Explorationists who are concerned with refining these quantities should find useful information in the following chapters.

In the past, velocity control was sparse for several reasons. Velocity processing not only was computationally costly but also involved some difficult, time-consuming interpretational effort. The results could not be quickly and conveniently displayed or digested. With increased computational efficiencies and expanded online and interactive capabilities, interpreters may now balance the cost of estimating velocity information continuously against the cost of not doing so. The tools now available can satisfy a geophysicist's needs to review his velocity data rapidly and make interpretational choices. While we do not anticipate that the digital computer will be able to interpret seismic data automatically, we expect it increasingly to assist the interpreter in his task.

There is little doubt that, in the future, geophysicists will call increasingly upon the principles of seismic wave theory to exploit seismic reflection measurements more fully. In addition to concentrating solely upon primary reflections and diffractions (as is done here), they will eventually include redundant information about the subsurface data such as, perhaps, certain multiple reflections that are source-generated. A high degree of redundancy is available in the seismic reflection method. This will likely be increasingly exploited to reduce the uncertainties inherent in subsurface parameter values computed from only a selected set of primary reflections or diffractions. Apart from the familiar exploitation of redundancy in CDP gathers, the inverse methods considered here make little use of redundant information in seismograms. We do show, however, that one and the same subsurface velocity model can be recovered as well from *migration velocities* used with time-migrated reflections as it can from *stacking velocities* combined with stacked reflections. We hope that our approach provides sufficient stimulus for some readers to consider taking advantage of other seismic information (e.g., multiples, shear waves) now considered extraneous to solving the inversion problem.

There are many good books dealing with various aspects of elastic or acoustic wave theory and ray theory (Ewing et al, 1957; Brekhovskikh, 1960; Cagniard, 1962; Grant and West, 1965; White, 1965; Bath, 1968; Cerveny and Ravindra, 1971; Gassmann, 1972; Stavroudis, 1972; Officer, 1974; Claerbout, 1976; and Cerveny et al, 1977). They are concerned primarily with solving *forward seismic problems*, i.e., computing wavefields for acoustic or elastic models of varying degree of complexity. Most of these authors rarely use the terminology of today's exploration seismologists, nor do they consider present-day field and processing techniques.

While most work in the past has been concerned with explaining physical measurements in terms of a specified experiment for a known model, solutions to such forward seismic problems can themselves be most useful for solving inverse seismic problems. The simulation capability of digital computers can be called upon to carry out an iterative series of searches for a model that leads to improved explanations of observed seismic traces or processed data.

Such iterative searching techniques, however, are as yet rarely implemented routinely in seismic exploration because they put high demands on both the capacity and computational speed of digital computers.

This monograph has been written with the a priori aim to be practice-oriented and to provide first-order accurate, economical solutions. Essentially, we use ray theory in *reverse gear*; that is, we concentrate upon providing noniterative inverse solutions based on ray theoretical considerations only. The inverse algorithms we discuss rely on analytical formulas which make it possible to express certain useful surface measurements (e.g., NMO velocities and small-aperture migration velocities) in terms of subsurface parameters. The algorithms we describe, however, can be integrated into more complex inverse seismic modeling methods based on the complete wave equation. Such methods will certainly become increasingly useful for analyzing specific wave propagation problems. An additional advantage of the inverse traveltime algorithms we discuss is that they rely to a large extent on routine seismic processing techniques. We exploit advantages offered by the CDP technique and rely upon stacking and migration velocity analyses as currently used for providing optimum stacks and time migrations.

Our algorithms cannot utilize any information from wavelets other than whatever contributes to defining *reflection times*. Therefore, any process (e.g., wavelet contraction, static correction, or noise suppression) which contributes to a better recognition of reflection times can be considered as being an integral part of the inverse methods.

Processes of subsurface image building based on the complete wave equation can essentially be viewed as processes of *backward wave propagation* or *downward continuation* (a progressive lowering of the datum plane to observe upward traveling waves is equivalent to propagation of these waves backward in time). Our ray-theoretical inverse traveltime methods are based on the same idea. They, too, can be viewed in terms of downward continuation. We attempt to make waves recorded at the earth's surface retreat to their points of reflection or diffraction. We thus use these waves as a "vehicle" to transport and to transform measurements obtained at the surface into desired subsurface parameters. The idea of using downward continuation to solve an inverse problem in seismic exploration is not new. It was successfully exploited in early refraction prospecting (Thornburgh, 1930) and is thus almost as old as seismic exploration itself.

2 Subsurface models

Three factors influence the accuracy with which subsurface parameters can be obtained—*measurement accuracy, aptness of the model,* and *completeness of the theory*. In exploration seismology, as in all other geophysical methods concerned with solving inverse problems, there exists a dualism between observation and theory which is fundamental. This dualism is embodied, in particular, in the notion of *subsurface models*. If the agreement between model and reality is poor, the model must be altered. Ambiguities are resolved by bringing in new information and altering the model. New ideas then take the form of new models. The complexity of the model depends on the amount and types of parameters one wants to extract. Though one has to acknowledge the likely existence of a complex geology in a seismic survey region or prospect area, one must at the same time allow for certain simplifying assumptions regarding lithology and structure, at least within the spread length of seismic receivers (which can be up to 5 km).

For our algorithms to work successfully without too many complications, we permit only smoothly curved first-order interfaces separating layers assumed to be continuous within a *local model*. In fact, we shall usually assume layers to be of constant velocity. A local model may be part of a larger *regional model* that need not be subjected to these restrictions. We also assume, in most of the work, that the radii of curvature of reflecting interfaces are much larger than the wavelengths of waves impinging upon them because, with this assumption, an interface will act predominantly as a reflector rather than a diffractor. We can thus associate propagating energy with (specular) raypaths. Specular reflection requires that the reflecting interface be smooth to within a small fraction of a wavelength. (What appears smooth to a low-frequency wave is not necessarily smooth for a high-frequency wave.)

In the seismic ray method, a first-order interface is a surface across which at least one of the generally smoothly changing Lamé parameters λ and μ or density ρ has a first-order discontinuity. A surface across which the lowest discontinuous spatial derivative of λ, μ, or ρ is of nth order, one refers to as an interface of $n + 1$th order (Cerveny and Ravindra, 1971). An elastic *inhomogeneous medium* in the sense used here is a medium characterized by smooth *spatially variable* parameters λ, μ, ρ. A *homogeneous medium* strictly requires λ, μ, and ρ to be constant.

Inhomogeneous media do not scatter energy for very high-frequency (i.e., short-wavelength) signals. Such energy can be returned only from first- and second-order interfaces. Homogeneous media do not scatter any energy at all, not even low-frequency signals. In this monograph, we will consider, in most cases, layered media consisting of discrete homogeneous media separated by smooth first-order interfaces.

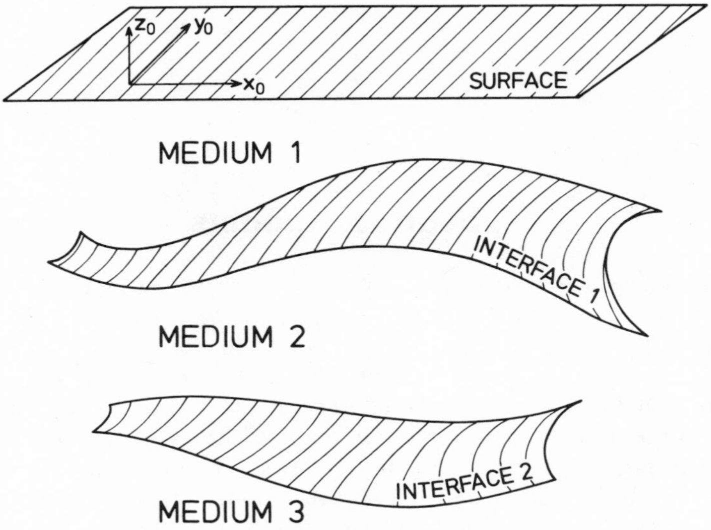

FIG. 2-1. 3-D subsurface model featuring three velocity layers separated by smoothly curved first-order interfaces.

The three elastic parameters λ, μ, and ρ govern the velocities of both *compressional* waves (*P*-waves) and *shear* waves (*S*-waves) in an isotropic medium:

$$V_P = \sqrt{\frac{\lambda + 2\,\mu}{\rho}}\,,$$

$$V_S = \sqrt{\frac{\mu}{\rho}}\,.$$

When $\mu = 0$, no shear waves can be transmitted; the medium is called an acoustic medium. In the seismic reflection method, acoustic wave theory based on the scalar wave equation is frequently considered because the influence of shear waves is (for almost vertically traveling waves) often small and negligible; as a result, computations are much simplified. (It should be mentioned, however, that neglect of shear waves generally unduly constrains interpretations of seismic data.) In inhomogeneous media, both *P*-wave and *S*-wave velocities may vary from point to point. For these media, it is useful to introduce the notion of isovelocity surfaces. Such a surface is the locus of points having the same local velocity in the continuously changing velocity medium. Quite often, reflecting interfaces are much steeper than isovelocity surfaces. Elastic layer properties and densities depend on both geologic age and depth, past and present; thus the same is valid for subsurface velocities. "Geologic age" naturally implies that older probably means deeper and thus age affects more compaction. Because of our preoccupation with kinematic problems, in most examples, the terms "homogeneous" and "inhomogeneous" are confined to the description of the

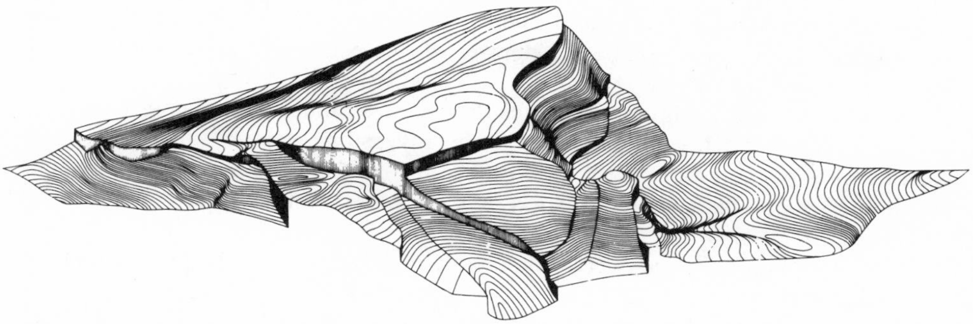

FIG. 2-2. Isometric view of real reflecting depth horizon interpreted from field data.

velocity field only. In fact, we will call a layer homogeneous if V_P (and not necessarily ρ or V_S) is a constant.

Models used below, in the study of rays and traveltimes, cannot account for the large number of interfaces found in reality and suggested by well logs. For practical reasons, we limit models to those containing only relatively few layers that most likely influence traveltimes. A typical model, used frequently later in this work, is illustrated in Figure 2-1. Its top surface can be taken to represent the earth's surface or some common datum plane to which all surface measurements are reduced. At best, the model can provide no more than an approximation to the true (more complex) local velocity distribution, and the selected first-order velocity boundaries just happen to represent a few (but, hopefully, the most influential) of the many reflecting interfaces that may exist.

Figure 2-2 shows an *isometric plot* of the depth of a selected mapped reflecting horizon over an actual prospect area. It was derived from traveltime measurements with one of the methods described later. The derived shape resulted from compositing many locally computed, constant-velocity layer models that included the selected horizon.

Traveltime inversion algorithms based upon thick constant-velocity layer models can, at best, provide velocities that are only average values over packs of thinner velocity layers. The so-derived velocities mostly agree with those averaged from well logs. However, when thin layers of very different physical properties alternate within a sequence (such as coal seams in shales), the integrated velocity log may often yield a higher average velocity than is derived from seismic reflections. Such observations have been explained in studies of transversely isotropic media, media in which velocity in the vertical direction differs from that in other directions (Postma, 1955; Uhrig and Van Melle, 1955; Helbig, 1965). Also, O'Doherty and Anstey (1971) have shown that with increasingly thin layering, wavelets transmitted through the layers may, due to interference, progressively lose their high-frequency content and have their peak amplitudes delayed. This delay can account for mis-ties of traveltimes observed in surface and subsurface recordings.

Nevertheless, even though the influence of intermediate reflectors is ignored in constructed models, computed depths of reflecting horizons often come close to their actual values.

Typically, a balance must be struck between simplicity and complexity of a model; use of too few interfaces for the inversion process can result in an insufficient approximation to the real object. In contrast, computing subsurface parameters for too many layers results in an increased uncertainty of all derived values.

If only the kinematic aspects of a wave propagation problem need to be modeled, a stack of thin layers can usually be replaced by one thick homogeneous layer of an appropriate average velocity. This averaging is, however, not acceptable for explaining seismic amplitudes. This is easily demonstrated by simulating the amplitudes of real seismic traces using synthetic seismograms. The dynamic properties of a propagating pulse are directly related to the nature of thin layering within the real earth (O'Doherty and Anstey, 1971). In this sense, amplitude is a much more sensitive parameter than is traveltime.

The constant-velocity layer models considered often satisfactorily approximate the real velocity distribution in the earth. Nevertheless, we must remain aware of various anomalous overburden configurations (such as those related to reefs, faults, flexures, unconformities, acoustic lenses, and strong lateral velocity gradients) which are not characterized by our models. Such real subsurface configurations, particularly when near the earth's surface, can be in severe conflict with the selected models for which our theory is developed. If a subsurface model agreed totally with the true geology, and surface measurements were free from uncertainties (neither condition is ever satisfied), then computed velocities would represent *seismic wave velocities*. Even though these conditions cannot be met, one may still want to interpret the derived velocities in terms of material distributions at depth as well as other velocity-affecting parameters like porosity, grain size, sand-shale ratio, pressure, and brine, oil, and gas content.

3 Seismic wave velocities

This brief chapter is included prior to embarking on the central theme so as to remind us of the physical rock properties which ultimately govern all seismic wave propagation.

Seismic wave velocity pertains to the speed of a seismic disturbance propagating through a material medium; it is a physical property of the medium. Particle velocity, on the other hand, refers to the actual motion of a local portion of the medium; it is a function of the disturbance of the medium rather than a property of the medium.

Whereas seismic wave velocities have dimensions of thousands of meters per second, particle velocities are typically in the order of millionths of a meter per second. We will not discuss particle velocity further in this work as it relates more to the dynamics than kinematics (i.e., amplitude rather than traveltime) of wave propagation.

Many field, laboratory, and theoretical studies have been devoted to the problem of determining seismic wave velocities and establishing empirical and analytical laws between these velocities and the various parameters that influence them (Faust, 1951; Gassmann, 1951; Wyllie et al, 1956, 1958; Press, 1966; Gardner et al, 1974a; Domenico, 1974; Toksöz et al, 1976; and Timur, 1977).

Temperature and pressure (which are dependent mainly on depth), as well as lithology, grain packing, porosity, and cementation all affect seismic wave velocity. Variations in lithology and the fluid and gas content of a porous rock can be important sources of strong velocity variation. Likewise, microfracturing can cause major reductions in material velocity.

The dependencies of velocity upon these many factors are not easily measured. Measurement of seismic wave velocities under real in-situ conditions is both difficult and expensive. Moreover, in the laboratory, it is not feasible to use sample sizes comparable to seismic wavelengths.

Fortunately, dispersion, the dependence of material velocities on frequencies, is small in the acoustic frequency range from 10^{-1} to 10^{7} Hz. Thus, most laboratory measurements are sufficiently independent of the method by which they are obtained. However, each rock type described lithologically has a large range of velocity values (sedimentary more so than igneous rocks). Only for shales does velocity increase more or less smoothly with depth, following a local compaction law within a thick layer.

Because velocity ranges for different rock types overlap to a large extent, it will be impossible to arrive at a unique interpretation of rock distributions from seismic reflection time measurements alone. Certain ambiguities, however, can be lessened by measuring dynamic reflection characteristics such as amplitude, signal strength, frequency content, and polarity. The value of using such characteristics has been demonstrated to some extent by the success of bright

Table 3-1. *P*-Wave velocities.

Material	V_P (km/sec)	Material	V_P (km/sec)
Alluvium	0.5–3.5	Quartz	3.8–5.2
Clay	1.1–2.5	Sandstone	1.4–5.1
Loam	0.8–1.8	Limestone	1.7–7.0
Sand	0.2–2.0	Salt	4.4–6.5
Weathered Layer	0.3–0.9	Caprock	3.5–5.5
Sand and Gravel	0.3–1.7	Anhydrite	4.0–5.0
Dolomite	3.5–6.9	Gypsum	2.0–3.5
Marble	3.7–7.3	Gneiss	2.0–5.3
Basalt	3.0–6.4	Water	1.47–1.56

spot (Gardner et al, 1974a; Hammond, 1974; Backus and Chen, 1975; and Dobrin, 1976) and pseudosonic log technology (Lavergne and Willm, 1977) as well as by other methods (Boisse, 1978).

Bright spot methods can provide direct, mutually compatible indicators for the detection of partially gas-saturated sands. True amplitude and velocity anomalies are analyzed jointly in bright spot work. The success of the bright spot and other techniques emphasizes that the traveltime inversion methods discussed here solve only one part of the *general seismic inversion problem*.

Table 3-1 shows representative laboratory *P*-wave velocities for various rock types (Clark, 1966). The overlap of velocity ranges should be a reminder that the "velocity problems" faced by seismic interpreters are, by no means, solved once interval velocities have been computed. The velocities that we can derive, however, should be sufficiently accurate to provide time-to-depth conversions using classical methods or modern depth migration techniques.

Readers who wish to gain further insight regarding the various factors affecting velocity in solid or porous rock and who wish to know about the use of seismic velocities for lithological studies will find little relevant information in this monograph. They must turn elsewhere (e.g., Anstey, 1977, and Domenico and Crowe, 1978). Our preoccupation is with the mathematics of computing interval velocities from traveltimes and of using these velocities for various processing purposes (migration, in particular).

4 Ray theory

Suppose that the layers of the model in Figure 2-1 are inhomogeneous and isotropic; then both *P*- and *S*-wave velocities are functions of space coordinates but are independent of the direction of wave propagation. (*Anisotropic* media, in which velocities depend on the direction of wave propagation at each point, are considered only briefly in Chapter 9.) Isotropic media in solid contact with one another can host two independent types of waves: One propagates with the local velocity of compressional waves, the other with the local velocity of shear waves.

The wavefront of an *elementary wave* can be described by the moving surface $t = \tau\ (x_0, y_0, z_0)$ defined with respect to some arbitrary right-hand $[x_0, y_0, z_0]$ coordinate system (Figure 2-1). The function $\tau(x_0, y_0, z_0)$ is known as the *phase function*. It satisfies the *eikonal equation*, a first-order partial differential equation of fundamental importance, for it leads directly to the concept of rays (Officer, 1974). Trajectories orthogonal to moving wavefronts are designated as rays. These rays can as well be obtained as the extremals of Fermat's functional, which represents the traveltime between any two designated points. Strictly speaking, the concept of a ray is appropriate only if energy propagates along a particular path, as in the case of a geometrical optics solution to the wave equation (Kline and Kay, 1965). Such solutions are obtained from the wave equation when the propagating pulse has wavelengths that are small in comparison with the structural dimensions of a model (e.g., radii of curvature of reflecting interfaces) and when spatial changes of the Lamé parameters and density are small over the distance of a wavelength (Officer, 1974; Cerveny and Ravindra, 1971).

The ray-theoretical approach used here deals only with the computation of traveltimes and raypaths. This classic ray theory, however, is an integral part of the more general, so-called *asymptotic ray method*, which provides approximate dynamic solutions to the elastic (or acoustic) wave equation for the entire wave field. When used in seismology, the more general approach has been confined largely to the study of noninterfering seismic body or head waves in generally thick, but otherwise complex, layered media with smooth interfaces.

Ray theory is particularly useful in simplifying otherwise complicated problems in wave propagation. Its application has been primarily in forward modeling studies, in which traveltimes are predicted for experiments in known models. There also exists an *inverse ray theory* which can provide exact recursive (noniterative) solutions to various inverse traveltime problems. Included as part of this inverse ray theory are algorithms for computing interval velocities and performing time-to-depth migrations (described later). Inverse ray theory can be looked upon as the ray-theoretical counterpart for *inverse wave equation theory*. Included in this latter category are the various noniterative methods,

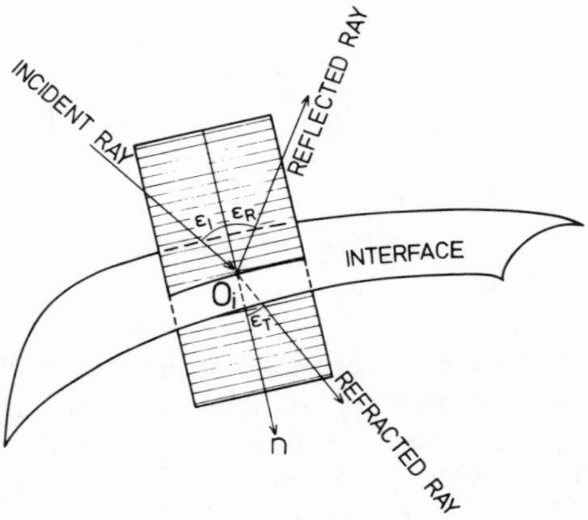

FIG. 4-1. Rays in the plane of incidence at a 3-D interface separating two constant-velocity layers in solid contact.

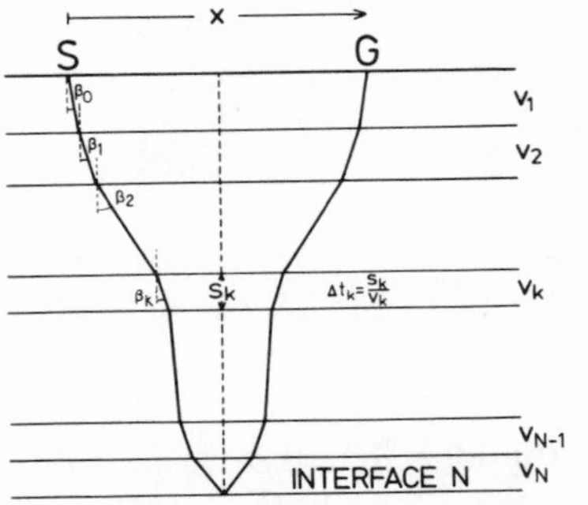

FIG. 4-2. Plane horizontal isovelocity layers.

based on finite difference and integration schemes (Claerbout, 1971, 1976; Walter and Peterson, 1976; Schneider, 1978; Stolt, 1978) that aim at transforming a concentration of observed seismic traces directly to an image of the earth's subsurface. As classic ray theory based on the eikonal equation involves only traveltimes and ray trajectories, it can be used only to solve the kinematic part of a wave propagation problem. This apparent disadvantage is, however, compensated by the fact that it can provide exact kinematic solutions in complex modeling problems (both forward and inverse) not easily treated by general wave theory. We hope that the ray-theoretical inverse solutions provided here may not only complement certain wave-theoretical solutions, but contribute to a better understanding of such inverse wave-equation problems as seismic imaging (or migration), of considerable importance in seismic exploration today. As indicated above, most of the inverse ray theoretical solutions given here can also be viewed as processes of downward continuation—a conceptual approach that has become very successful in seismic imaging.

Equations that describe wavefronts or ray trajectories (i.e., eikonal equation or ray tracing systems) can be derived from various wave equations, partial differential equations that govern, for instance, particle displacement, particle velocity, or pressure in media and at interfaces. When the earth is modeled as a stack of homogeneous layers or blocks, ray theory requires no more than two simple conditions to be satisfied. One is that rays are straight trajectories within each layer. The other is that they obey Snell's law at interfaces. The first of the two conditions is actually a necessary consequence of Snell's law which can be expressed as follows:

$$\frac{\sin|\varepsilon_I|}{v_I} = \frac{\sin|\varepsilon_T|}{v_T} = \frac{\sin|\varepsilon_R|}{v_R},$$

where ε_I is the incidence angle, ε_T the refraction angle, and ε_R the reflection angle (Figure 4-1).

The incident ray, refracted ray, and the reflected ray, together with the interface normal vector $\mathbf{n}$, lie within a common plane—the plane of incidence. v_I and v_R are velocities on the incident side, and v_T is the velocity on the refracted side of the interface. These velocities are the speeds of propagation for either compressional or shear waves. For most seismic applications, they are compressional wave velocities, and $v_R = v_I$.

The basic aim here is to develop recursive traveltime inversion methods for 3-D models like that in Figure 2-1. Such models we call "3-D isovelocity layer models." These general models include as a special and important case the plane horizontal, isovelocity layer model of Figure 4-2. Numerous authors (Krey, 1951; Dix, 1955; Schmitt, 1966; Taner and Koehler, 1969; Robinson, 1970a; Al-Chalabi, 1973, 1974; Shah and Levin, 1973; and Marschall, 1975) have investigated various traveltime aspects of this simple model that are of general interest to exploration seismologists. In the next section, we give a tutorial treatment of this plane horizontal layer model to demonstrate to newcomers in seismic exploration the manner in which reflection time curves conceal desired subsurface information. We shall see some of the mathematical complexity of reflection time curves and show where simplifying assumptions can be made, for instance, for purposes of computing interval velocities from stacking velocities.

Ray tracing in 3-D isovelocity layers is not as simple as in the plane horizontally layered case. We shall devote the second section of this chapter to this particular subject and later make use of results in the solution of various forward and inverse traveltime problems. Wavefront curvature laws provide the key to solution of most problems treated in the remainder of the monograph. These laws are discussed in the third section. In the fourth section, we show how these laws can help solve, in an iterative fashion, the problem of tracing rays between specified end points. The role of wavefront curvatures in geometrical spreading and the close relationship between geometrical spreading and NMO are discussed in the fifth section. That section concludes with a brief review of the asymptotic ray method.

4.1 Plane horizontal layers

In this section, we shall derive an expression for the traveltime squared as a Taylor series expansion in powers of the square of the shot-receiver distance. Beyond the second term (i.e., the hyperbolic approximation), the expressions for the coefficients of the series become complicated. We include their derivation primarily to demonstrate those circumstances (i.e., layering configurations and source-to-receiver distances) for which the hyperbolic approximation is inadequate. In practice, little use can be made of the higher-order terms because of complications that arise when the actual subsurface does not consist simply of horizontal layers. Moreover, noises that contaminate measurements of traveltime in real data render the matching of the higher-order terms highly uncertain. The Taylor series expansion derived in this section does not readily generalize for models in which layers are not simply plane and horizontal.

Let a source S and receiver G (Figure 4-2) be separated by the distance x on the earth's surface. The traveltime of a primary wave reflected from the Nth interface between S and G is designated as $T(x)$. It can be expressed in the following parametric form (each layer is counted twice, i.e., $v_k = v_{2N+1-k}$ and $\Delta t_k = \Delta t_{2N+1-k}$)

$$x(p) = \sum_{k=1}^{2N} \frac{v_k^2 \Delta t_k p}{\sqrt{1 - v_k^2 p^2}}, \tag{4.1}$$

and

$$T(p) = \sum_{k=1}^{2N} \frac{\Delta t_p}{\sqrt{1 - v_k^2 p^2}}, \tag{4.2}$$

where the ray parameter p is given by $p = \sin \beta_k / v_k$. β_k is the angle of the ray from vertical, v_k is the velocity, and Δt_k is the one-way vertical time in layer k.

For the reflection from the base of the first layer, we can eliminate p and obtain the familiar hyperbolic traveltime curve

$$T^2(x) = T^2(0) + \frac{x^2}{v_1^2}.$$

Thus, the velocity v_1 above the uppermost reflector is immediately computable from the hyperbola since $1/v_1^2$ corresponds to the slope of the straight line that results when plotting T^2 against x^2.

$T(x)$, a symmetric curve for the multilayer case as well, can be written in the following form (Taner and Koehler, 1969; Marschall, 1975):

$$T^2(x) = C_0 + C_1 x^2 + C_2 x^4 + C_3 x^6 + \dots \tag{4.3}$$

with $C_0 = T^2(0)$.

The coefficients C_1, C_2, ... can be determined by expressing T^2, x^2, x^4, x^6, etc., as power series in p. Then, by substituting these power series expressions into equation (4.3), we can find the desired coefficients C_1, C_2, ... and comparing like-power terms for $|p| < \min(v_k^{-1})$, we can expand $x(p)$ into the absolutely convergent Taylor series

$$x = \sum_{k=1}^{2N} v_k^2 \Delta t_k p \left(1 + \frac{1}{2} p^2 v_k^2 + \frac{1 \cdot 3}{2 \cdot 4} p^4 v_k^4 + \frac{1 \cdot 3 \cdot 5}{2 \cdot 4 \cdot 6} p^6 v_k^6 + \dots,\right) \tag{4.4}$$

or concisely

$$x = \sum_{j=1}^{\infty} A_j^{(1)} p^{2j-1}, \tag{4.5}$$

where

$$A_j^{(1)} \equiv \sum_{k=1}^{2N} v_k^{2j} \Delta t_k \left(\frac{2j}{2j-1} \frac{(2j)!}{4^j j! j!}\right) = J_j V_{(2j)} T(0), \tag{4.6}$$

$$J_j \equiv \frac{2j}{2j-1} \frac{(2j)!}{4^j j! j!}; \left(J_1 = 1, J_2 = \frac{1}{2}, J_3 = \frac{1 \cdot 3}{2 \cdot 4}, \dots\right), \tag{4.7}$$

$$\frac{1}{2} V_{(2j)} T(0) \equiv \sum_{k=1}^{N} v_k^{2j} \Delta t_k . \tag{4.8}$$

From equation (4.5), we can derive absolutely convergent series for x^2, x^4, x^6, x^8, etc. For x^2, we get

$$x^2(p) = \sum_{j=1}^{\infty} A_j^{(2)} p^{2j}, \tag{4.9}$$

where

$$A_j^{(2)} = \sum_{\mu=1}^{j} A_\mu^{(1)} A_{j-\mu+1}^{(1)}, \tag{4.10}$$

with

$$A_1^{(2)} = A_1^{(1)} A_1^{(1)},$$
$$A_2^{(2)} = A_1^{(1)} A_2^{(1)} + A_2^{(1)} A_1^{(1)},$$
$$\cdots\cdots\cdots\cdots\cdots\cdots$$

Generally, one obtains for x^{2n}:

$$x^{2n}(p) = \sum_{j=1}^{\infty} A_j^{(2n)} p^{2j+2n-2}, \tag{4.11}$$

where

$$A_j^{(2n)} = \sum_{\mu=1}^{j} A_\mu^{(2n-2)} A_{j-\mu+1}^{(2)} . \tag{4.12}$$

In a similar way as for x^2, x^4, etc., one can find power series expansions for T and T^2:

$$T = \sum_{j=0}^{\infty} B_j^{(1)} p^{2j}, \tag{4.13}$$

where

$$B_0^{(1)} = T(0), \tag{4.14}$$

and

$$B_j^{(1)} = J_{j+1} V_{(2j)} T(0).$$

J_{j+1} and $V_{(2j)}$ are the same as in equations (4.7) and (4.8).

From equation (4.13), it follows that

$$T^2 = \sum_{j=0}^{\infty} p^{2j} B_j^{(2)}, \tag{4.15}$$

with

$$B_j^{(2)} = \sum_{\mu=0}^{j} B_\mu^{(1)} B_{j-\mu}^{(1)} . \tag{4.16}$$

One can now compute the coefficients C_i ($i = 1, 2, \ldots$) by substituting equation (4.11) for $n = 1, 2, 3$ and equation (4.15) into equation (4.3). Comparing coefficients for the same powers of p provides:

$$C_0 = B_0^{(2)},$$
$$C_1 = B_1^{(2)} / A_1^{(2)},$$

and

$$C_2 = [\, B_2^{(2)} - C_1 A_2^{(2)}] / A_1^{(4)},$$

or generally,

$$C_n = \left[B_n^{(2)} - \sum_{j=1}^{n-1} C_{n-j} A_{j+1}^{2(n-j)} \right] \Big/ A_1^{(2n)};\ n = 2, 3, \ldots. \tag{4.17}$$

By considering equations (4.7) and (4.8), one finally obtains for C_0 to C_4

$$C_0 = T_0^2 = T^2(0),$$
$$C_1 = \frac{1}{V_{(2)}},$$
$$C_2 = \frac{V_{(2)}^2 - V_{(4)}}{4\, V_{(2)}^4\, T_0^2},$$
$$C_3 = \frac{2\, V_{(4)}^2 - V_{(6)} V_{(2)} - V_{(4)} V_{(2)}^2}{8\, V_{(2)}^7\, T_0^4},$$
$$C_4 = \frac{-24 V_{(4)}^3 - 5\, V_{(8)} V_{(2)}^2 - 4\, V_{(6)} V_{(2)}^3 + 9\, V_{(4)}^2 V_{(2)}^2 + 24\, V_{(6)} V_{(4)} V_{(2)}}{64\, V_{(2)}^{10}\, T_0^6}.$$

More coefficients of the series (4.3) are discussed by Marschall (1975).

The quantities $V_{(m)}^{1/m}$ designate weighted averages of velocity in the following form:

$$V_{(m)}^{1/m} = \left[\frac{2}{T(0)} \sum_{i=1}^{N} v_i^m \Delta t_i \right]^{1/m}.$$

In particular, we have for $m = 2$

$$V_{(2)}^{1/2} = \left[\frac{2}{T(0)} \sum_{i=1}^{N} v_i^2 \Delta t_i \right]^{1/2} = V_{\mathrm{RMS}}. \tag{4.18}$$

V_{RMS} is generally referred to as the RMS (root mean square) velocity. $1/V_{\mathrm{RMS}}^2$ is the slope of the $T^2 - x^2$ curve at $x^2 = 0$.

From equation (4.18), the reader will deduce that V_{RMS} differs from the average velocity

$$V_A = \frac{2}{T(0)} \sum_{i=1}^{N} v_i \Delta t_i = \frac{2z}{T(0)}$$

(Krey, 1951, p. 479–484; Dix, 1955), which relates vertical traveltime to depth z in a horizontally layered medium. From the above derivation, we have the squared time $T^2(x)$ expanded entirely in terms of various time-weighted average velocity quantities along the *vertical* travel path that results for $x = 0$. It could

Table 4-1. Traveltimes for truncated Taylor series expansions.

x_i [km]	0	1	2	3	4	5	6	7	8	9	10	C_i
$T(x_i)$ [msec]	4000	4009	4037	4083	4146	4225	4321	4431	4554	4690	4838	2
	4000	4009	4037	4082	4145	4223	4316	4423	4541	4670	4808	3
	4000	4009	4037	4082	4145	4223	4316	4423	4542	4672	4811	4
	4000	4009	4037	4082	4145	4223	4316	4423	4542	4672	4811	5
	4000	4009	4037	4082	4145	4223	4316	4423	4540	4666	4795	6
	4000	4009	4037	4082	4145	4223	4316	4423	4541	4669	4805	7

as well be expanded in terms of parameters along any other oblique reflected ray (Shah and Levin, 1973). One would then find that the hyperbola (with apex at $x = 0$) which provides the best second-order approximation for $T(x)$ in the vicinity of x will be defined by the RMS velocity computed along the *slanting* reflected raypath between S and G.

As $T(0)$ and all velocity averages are independent of the ordering of the layered sequence, layers can be interchanged arbitrarily above a given reflector without affecting the reflection time curves of that reflector or of deeper ones. For small values of x, the function $T^2(x)$ is determined predominantly by the RMS velocity. RMS velocities and two-way normal times can thus be estimated from the traveltime curves of all primary reflections and then equation (4.18) can be solved to recover the interval velocities v_i ($i = 1, \ldots, N$). Interesting also, if $2N-1$ coefficients of the series for $T^2(x)$ could be established from the Nth primary reflection time curve, then it is theoretically possible to compute all interval velocities and thicknesses of layers (but not their order) in the overburden from this one reflection time curve alone. However, large uncertainties in measurements from actual data render the *redundant information* about the overburden contained in the higher-order terms of the series for $T^2(x)$ generally of little practical value.

We would probably deter most geophysicists from reading on if we extended the mathematical analysis of $T^2(x)$ with discussions of convergence and other fine points. We will, therefore, refrain from proofs and only mention some interesting points; for instance, the first four coefficients have alternating signs ($C_1 > 0$; $C_2 < 0$; $C_3 > 0$; $C_4 < 0$) and the series (4.3) is rapidly convergent if the ratio of shot-receiver distance to depth is small (Taner and Koehler, 1969; Al-Chalabi, 1973; Marschall, 1975). Strong oscillations occur when the ratio is high. The importance of terms higher than the second-order term in $T^2(x)$ is revealed to some extent in the following example, representative of many practical situations.

A four-layer model (Marschall, 1975) is chosen with velocities $v_1 = 2000$ m/sec, $v_2 = 3000$ m/sec, $v_3 = 4000$ m/sec, and $v_4 = 5000$ m/sec. The two-way reflection times (for $x = 0$) to the lower interfaces of each layer are 1.0 sec, 2.0 sec, 3.0 sec, and 4.0 sec, respectively. Let us truncate the series expansion for $T^2(x)$ before the term with $C_2, C_3, \ldots,$ up to C_7, respectively. Table 4-1 shows the times $T(x_i)$ in msec to the fourth interface for $x_i = 0, 1, \ldots, 10$ km, for these truncated series. The example demonstrates the negligible influence of terms higher than the second-order term for $x_i \leq 3$ km. Terms higher than C_2 also have no influence on the traveltime for 3 km $\leq x_i \leq$ 7 km. The term with C_3 becomes noticeable for 7 km $\leq x_i \leq$ 8 km (7 km just happens to be the depth of the horizon in question). Even for $x_i > 8$ km, the contribution of the higher-order terms to the accuracy of the curve is of little practical value. In solving inverse traveltime problems, it would be extremely difficult to establish these terms from $T^2(x)$ curves observed in conventional reflection seismic surveys.

4.1.1 Summary

It is hoped that this short review on reflection time curves for plane horizontal, isovelocity layers is sufficient to justify why terms in $T^2(x)$ higher than x^2 have little practical value for the theory of computing interval velocities in such models. This conclusion is valid up to rather large shot-geophone distances. Even for the plane parallel layer case, the information in the higher-order terms is already bound up in a mathematically complicated form. Rather than using the higher-order terms, one will generally try to suppress their influence by restricting the offset distance x to small values. Then the constant and x^2 terms dominate most of the traveltime curve.

Terms of higher order than x^2 in the power series expansion of a primary reflection for a 3-D isovelocity layer model would be extremely complex, and we have no desire to compute such terms. In practice, higher-order terms are dealt with numerically (by means of costly ray tracing) rather than analytically. Fortunately, even for very complex subsurface models, interval velocities can be recovered from a precise knowledge of just the zero- and second-order terms for each so-called primary CDP *reflection time curve* (see chapter 5). This will become evident later. It should be clear, therefore, why we put much emphasis on determining analytic expressions (in terms of model parameters) for only these terms. We will approach this task with the help of wavefront curvature laws. The particular method based on these laws may appeal more to the reader than will the approach reviewed in this section. Although our analytic approach, strictly limited to small offset distances and isovelocity layers, is less general than numerical approaches, it provides larger insights into traveltime behavior in complex media and offers the possibility of solving inverse problems in a noniterative manner.

4.2 Ray tracing

We now present a recursive algorithm for tracing a ray through the 3-D layered model of Figure 2-1, where layer velocities are constant. Models of this type often quite well approximate more complex velocity models where layer velocities are inhomogeneous. As ray theory demands no more than to have Snell's law satisfied at interfaces, all other considerations in connection with solving the ray-tracing problem are of a purely geometrical and computational nature. Vector and matrix algebra provide the most suitable mathematical tools.

Let us trace the ray that starts from point $\mathbf{O}_0$ in direction $\mathbf{e}_0$ through the

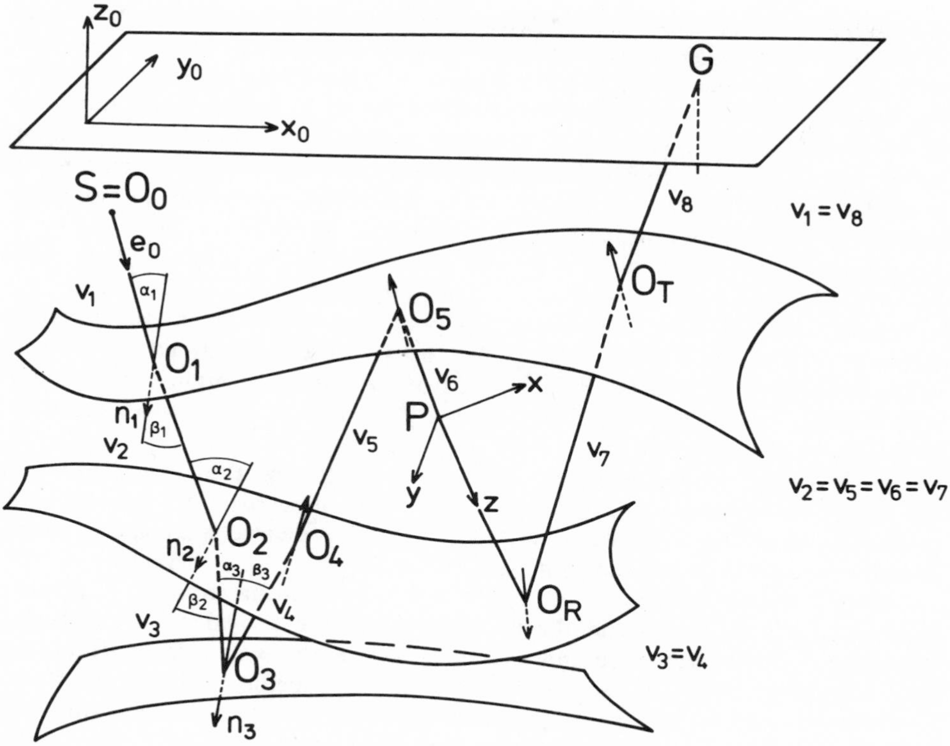

FIG. 4-3. 3-D isovelocity layer model featuring a ray between a source S and a receiver G. The moving $[x, y, z]$ frame is indicated at ray point P. Points $\mathbf{O}_1$, $\mathbf{O}_2$, and $\mathbf{O}_4$ are points of refraction ($\mathbf{O}_T$), and $\mathbf{O}_3$ **and** $\mathbf{O}_5$ are points of reflection ($\mathbf{O}_R$).

velocity model shown in Figure 4-3. For the forward modeling problem, we assume that all layer velocities and first-order interface shapes are known. Subsurface velocity boundaries are provided in the form of discrete points with respect to the $[x_0, y_0, z_0]$ system. Should interfaces be available in any other form, we have them sampled in this way. According to Snell's law, the initial ray direction $\mathbf{e}_0$ determines the entire ray through the model. There is no way of explicitly choosing $\mathbf{e}_0$ at $\mathbf{O}_0$ a priori in such a way that the ray passes through a specified end point. (In section 4.4, a "shooting method" is described for iterative determination of the ray trajectory between specified end points.)

Although it is easy to visualize the tracing of a ray through a system of isovelocity layers, it is yet another matter to formulate this process in adequate mathematical terms so that it can be programmed.

As one will soon appreciate, it is most helpful to work with a *moving* $[x, y, z]$ *right-hand coordinate system* as illustrated at wavefront point **P** in Figure 4-3. This coordinate frame accompanies the wavefront along a selected ray. By definition, the z-axis will point in the direction of the advancing wavefront. Thus, the x- and y-axes are confined to the plane tangent to the wavefront at **P**. We define the $[x, y, z]$ frame uniquely by requiring its x-axis to fall in the *plane of incidence* of the next interface. Obviously, two possibilities exist for the choice of the x-axis. We settle upon one of them within the first velocity layer, where the ray tracing is started. Henceforth the x-axis is determined in all layers. Having specified both the z- and x-axes, one selects the y-axis so that the resulting coordinate system is right-handed. Note that in order to specify the moving system within each constant-velocity layer, one needs to know the interface normal vector for the point where the ray encounters the next interface. Interface normal vectors point, by definition, in the direction *away* from the incident wave. In order to satisfy all definitions with regard to the moving system, it is obvious that it must translate within homogeneous layers. Then, at each point of reflection $\mathbf{O}_R$ or refraction $\mathbf{O}_T$, it is subjected to two successive rotations.

Let $[x_I, y_I, z_I]$ designate the moving $[x, y, z]$ system on the incident side and $[x_R, y_R, z_R]$ and $[x_T, y_T, z_T]$ the system on the reflected and refracted sides of an interface. Let $[\bar{x}_R, \bar{y}_R, \bar{z}_R]$ and $[\bar{x}_T, \bar{y}_T, \bar{z}_T]$ denote auxiliary systems, at interface points, which are not part of the moving frame. They are defined as follows: The $[\bar{x}_R, \bar{y}_R, \bar{z}_R]$ and $[\bar{x}_T, \bar{y}_T, \bar{z}_T]$ systems result from the $[x_I, y_I, z_I]$ system by a rotation around the y_I-axis so that the turned z_I-axis points in the direction of the reflected and refracted ray, respectively. The resulting $\bar{x}_R$- or $\bar{x}_T$-axis is not generally confined to the plane of incidence belonging to the next interface. The $[x_R, y_R, z_R]$ system is subsequently obtained from the $[\bar{x}_R, \bar{y}_R, \bar{z}_R]$ system by a further rotation around the $\bar{z}_R$-axis. Likewise, the $[x_T, y_T, z_T]$ system is obtained from the $[\bar{x}_T, \bar{y}_T, \bar{z}_T]$ system by a further rotation around the $\bar{z}_T$-axis. The rotation angle in both cases is designated as δ. It can be interpreted as the angle by which the plane of incidence at the considered interface point has to be rotated around the $\bar{z}_R$- or $\bar{z}_T$-axis in order to coincide with the plane of incidence at the next interface. The sign of δ is defined below.

It is clear now that the $[x_R, y_R, z_R]$ or $[x_T, y_T, z_T]$ system on the reflected or refracted side of an interface is parallel to the $[x_I, y_I, z_I]$ system on the incident side of the next encountered interface. To simplify considerations, let $[\bar{x}, \bar{y}, \bar{z}]$ designate either the $[\bar{x}_R, \bar{y}_R, \bar{z}_R]$ or $[\bar{x}_T, \bar{y}_T, \bar{z}_T]$ system, depending on whether reflection or refraction is considered.

Let $\mathbf{O}_i$ designate the intersection point of a ray with the ith interface that the ray encounters (Figure 4-3). $\mathbf{O}_i$ can be a point of reflection $\mathbf{O}_R$ or a point of refraction $\mathbf{O}_T$. Let us also assume that the coordinates of point $\mathbf{O}_i$ and the directional vectors $\mathbf{e}_{\bar{x}}, \mathbf{e}_{\bar{y}}, \mathbf{e}_{\bar{z}}$ defining the $[\bar{x}, \bar{y}, \bar{z}]$ system attached to it have already been established with respect to the $[x_0, y_0, z_0]$ system. Then, once we compute the respective quantities associated with $\mathbf{O}_{i+1}$ in terms of the known quantities at $\mathbf{O}_i$, much of the ray-tracing problem is solved.

In the following, many parameters or coordinate systems should have the subscript i. For simplicity of notation, however, this subscript is omitted wherever no ambiguities exist. Let s_i be the distance between the points $\mathbf{O}_{i-1}$ and $\mathbf{O}_i$, and let $\Delta t_i = s_i / v_i$ be the traveltime between the two points. v_i is the velocity along the ith raypath segment. Ray segments are numbered consecutively from the ray origin, starting with one.

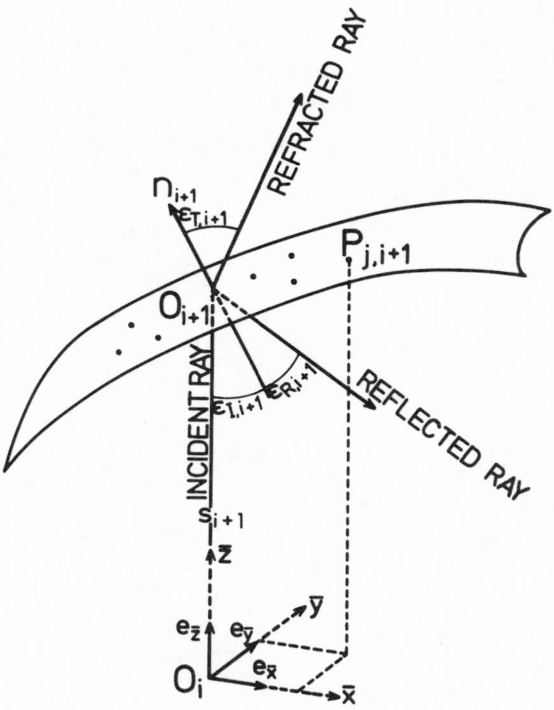

FIG. 4-4. Ray parameters considered when viewing the next interface (at $\mathbf{O}_{i+1}$) from within the local $[\bar{x}, \bar{y}, \bar{z}]$ system at $\mathbf{O}_i$. The signs of the angles are defined in step 4 of section 4.2.

The computational steps required to trace the ray from $\mathbf{O}_i$ to $\mathbf{O}_{i+1}$ will now be described. Some readers may find it advisable to skip some of the details in their first attempt to familiarize themselves with the basic principles of ray theory. They may quickly glance over the steps and refer back as the need arises.

Note that if an interface is a reflecting one, the $[\bar{x}, \bar{y}, \bar{z}]$ system represents the $[\bar{x}_R, \bar{y}_R, \bar{z}_R]$ system; otherwise, it represents the $[\bar{x}_T, \bar{y}_T, \bar{z}_T]$ system.

Step 1.—Transform the points of the $(i + 1)$th interface, assumed to be defined in the $[x_0, y_0, z_0]$ system, into the local $[\bar{x}, \bar{y}, \bar{z}]$ system at $\mathbf{O}_i$. Let $\mathbf{P}_{j,i+1}$ designate any point on the $(i + 1)$th interface. The coordinates of the vectors $\mathbf{O}_i, \mathbf{e}_{\bar{x}}, \mathbf{e}_{\bar{y}}, \mathbf{e}_{\bar{z}}$, and $\mathbf{P}_{j,i+1}$ are all defined in the $[x_0, y_0, z_0]$ system. Let $(\bar{x}_j, \bar{y}_j, \bar{z}_j)_{i+1}$ designate the coordinates of $\mathbf{P}_{j,i+1}$ with respect to the $[\bar{x}, \bar{y}, \bar{z}]$ system at $\mathbf{O}_i$. They can be obtained from the following equation (see Figure 4-4):

$$\bar{x}_{j,i+1}\,\mathbf{e}_{\bar{x}} + \bar{y}_{j,i+1}\,\mathbf{e}_{\bar{y}} + \bar{z}_{j,i+1}\,\mathbf{e}_{\bar{z}} = \mathbf{P}_{j,i+1} - \mathbf{O}_i. \tag{4.19}$$

Step 2.—Now let us assume that a sufficient number of points $\mathbf{P}_{j,i+1}$ have been transformed into the local $[\bar{x}, \bar{y}, \bar{z}]$ system at $\mathbf{O}_i$ to describe the $(i + 1)$th interface in the vicinity of the $\bar{z}$-axis. Provided there are no interface discontinuities (e.g., faults, truncations at unconformities), one can fit a second-order polynomial (for instance, by some least-square scheme) through those points $\mathbf{P}_{j,i+1}$, whose projections along the ray are close to the origin of the $\bar{x}$-$\bar{y}$ plane at $\mathbf{O}_i$. The resulting polynomial may be expressed as

$$\bar{z} = \bar{b}_{11}\bar{x}^2 + 2\bar{b}_{12}\bar{x}\bar{y} + \bar{b}_{22}\bar{y}^2 + \bar{c}_1\bar{x} + \bar{c}_2\bar{y} + \bar{s}. \tag{4.20}$$

In matrix notation, this can be written as:

$$\Phi(\bar{x}, \bar{y}, \bar{z}) = 0 = \bar{z} - (\bar{x}, \bar{y}) \begin{bmatrix} \bar{b}_{11} & \bar{b}_{12} \\ \bar{b}_{12} & \bar{b}_{22} \end{bmatrix} \begin{bmatrix} \bar{x} \\ \bar{y} \end{bmatrix} - (\bar{c}_1, \bar{c}_2) \begin{bmatrix} \bar{x} \\ \bar{y} \end{bmatrix} - \bar{s}, \tag{4.21}$$

or, more concisely,

$$\Phi(\bar{x}, \bar{y}, \bar{z}) = 0 = \bar{z} - \bar{\mathbf{X}}\bar{\mathbf{B}}\bar{\mathbf{X}}^T - \bar{\mathbf{C}}\bar{\mathbf{X}}^T - \bar{s}. \tag{4.22}$$

$\bar{\mathbf{B}}$ is a symmetric 2×2 matrix and $\bar{\mathbf{C}}$ and $\bar{\mathbf{X}}$ are row vectors. We shall consider all vectors as row vectors unless they are marked by the superscript T which denotes vector transpose. Clearly, then, the magnitude $\bar{s}$ is the desired distance s_{i+1} from $\mathbf{O}_i$ to $\mathbf{O}_{i+1}$. Though $\bar{\mathbf{B}}$ and $\bar{\mathbf{C}}$ define the basic features of the $(i + 1)$th interface, they refer to the local $[\bar{x}, \bar{y}, \bar{z}]$ system at $\mathbf{O}_i$. That is, both quantities should have the subscript i which has been suppressed. The matrix $\bar{\mathbf{B}}$ is not specifically required for ray tracing. It is, however, needed later on in the context of computing wavefront curvatures (see section 4.3 and Appendix A). If the interface points $\mathbf{P}_{j,i+1}$ fall close enough to the $\bar{z}$-axis, a first-order polynomial [substitute $\bar{b}_{ij} = 0$ in equation (4.21)] is sufficient to solve the ray-tracing problem.

Step 3.—Compute $\mathbf{O}_{i+1}$ and Δt_{i+1}. The point $\mathbf{O}_{i+1}$ can be expressed with respect to the $[x_0, y_0, z_0]$ system as:

$$\mathbf{O}_{i+1} = \mathbf{O}_i + s_{i+1}\mathbf{e}_{\bar{z}}. \tag{4.23}$$

The traveltime between $\mathbf{O}_i$ and $\mathbf{O}_{i+1}$ is $\Delta t_{i+1} = s_{i+1}/v_{i+1}$.

Step 4.—Compute the interface normal vector $\bar{\mathbf{n}}_{i+1}$ at $\mathbf{O}_{i+1}$ with respect to the local $[\bar{x}, \bar{y}, \bar{z}]$ system at $\mathbf{O}_i$. This vector points, by definition, in the direction of the gradient $\nabla\Phi = (\partial\Phi/\partial\bar{x}, \partial\Phi/\partial\bar{y}, \partial\Phi/\partial\bar{z})$ of $\Phi(\bar{x}, \bar{y}, \bar{z})$ at $\bar{x} = \bar{y} = 0$ (Figure 4-4).

It can be written as

$$\bar{\mathbf{n}}_{i+1} = \nabla\Phi/|\nabla\Phi| = \frac{1}{\sqrt{1 + \bar{c}_1^2 + \bar{c}_1^2}}(-\bar{c}_1, -\bar{c}_2, 1). \tag{4.24}$$

The incidence angle $\varepsilon_{I,i+1}$ at $\mathbf{O}_{i+1}$ is available from the following scalar product

$$\cos\varepsilon_{I,i+1} = \bar{\mathbf{n}}_{i+1} \cdot \bar{\mathbf{e}}_{\bar{z}} = 1/\sqrt{1 + \bar{c}_1^2 + \bar{c}_2^2}, \tag{4.25}$$

where $\bar{\mathbf{e}}_{\bar{z}} = (0,0,1)$ is the directional unit vector of the z-axis in the local $[\bar{x}, \bar{y}, \bar{z}]$ system at $\mathbf{O}_i$. The reflection angle $\varepsilon_{R,i+1}$ and the refraction angle $\varepsilon_{T,i+1}$ at $\mathbf{O}_{i+1}$ are available from Snell's law:

$$\frac{-\sin \varepsilon_{R,i+1}}{v_{R,i+1}} = \frac{\sin \varepsilon_{T,i+1}}{v_{T,i+1}} = \frac{\sin \varepsilon_{I,i+1}}{v_{I,i+1}}. \tag{4.26}$$

$v_{R,i+1}$ and $v_{T,i+1}$ pertain to layer velocities on the reflected and refracted side of $\mathbf{O}_{i+1}$, respectively. The minus sign in the first expression is due to the following sign convention which we use throughout this monograph. First, we require ε_I, ε_T, and ε_R to be acute, i.e., they satisfy the condition

$$-\frac{\pi}{2} < \varepsilon_I, \varepsilon_T, \varepsilon_R < +\frac{\pi}{2}.$$

Then we say that each angle ε_I, ε_T, or ε_R is positive if the z_I-, z_T-, or z_R-axis can be rotated by this acute ε-angle in the positive or negative direction of the interface normal vector in a clockwise manner of rotation around the $y_I = \bar{y}_T = \bar{y}_R$ axis when looking into the $+y$-direction. Thus in Figure 4-4, $\varepsilon_{I,i+1}$ and $\varepsilon_{T,i+1}$ would be negative, while $\varepsilon_{R,i+1}$ would be positive. Note that while ε_I and ε_T always have the same sign, ε_I and ε_R always have opposite signs.

Step 5.—Compute the angle δ_i between the planes of incidence at $\mathbf{O}_i$ and $\mathbf{O}_{i+1}$. This operation is simple if performed in the local $[\bar{x}, \bar{y}, \bar{z}]$ system attached to $\mathbf{O}_i$. The plane of incidence at $\mathbf{O}_i$ is defined as $\bar{y} = 0$. Its counterpart at $\mathbf{O}_{i+1}$ (again defined with respect to the $[\bar{x}, \bar{y}, \bar{z}]$ system at $\mathbf{O}_i$) is $\bar{c}_2\bar{x} - \bar{c}_1\bar{y} = 0$. The angle δ_i is hence available from $\tan \delta_i = +\bar{c}_2/\bar{c}_1$. When looking into the $+\bar{z}$-direction, a positive sign of δ_i requires a *clockwise* rotation of the plane of incidence at $\mathbf{O}_i$ to have it overlap with the plane of incidence at $\mathbf{O}_{i+1}$. Only after δ_i has been established can one compute the unit vectors $(\mathbf{e}_x, \mathbf{e}_y, \mathbf{e}_z)_i$ of the moving $[x, y, z]$ system on the reflected or refracted side at $\mathbf{O}_i$. This is achieved by the following transformation:

$$\begin{bmatrix} \mathbf{e}_x \\ \mathbf{e}_y \\ \mathbf{e}_z \end{bmatrix}_i = \begin{bmatrix} \cos \delta_i & \sin \delta_i & 0 \\ -\sin \delta_i & \cos \delta_i & 0 \\ 0 & 0 & 1 \end{bmatrix} \begin{bmatrix} \mathbf{e}_{\bar{x}} \\ \mathbf{e}_{\bar{y}} \\ \mathbf{e}_{\bar{z}} \end{bmatrix}_i . \tag{4.27}$$

Note that the unit vectors specifying the coordinate frames are expressed with respect to the $[x_0, y_0, z_0]$ system.

Step 6.—The directional unit vectors $(\mathbf{e}_{x_I}, \mathbf{e}_{y_I}, \mathbf{e}_{z_I})_{i+1}$ describing the $[x_I, y_I, z_I]_{i+1}$ frame at the incident side of $\mathbf{O}_{i+1}$ are equal to the unit vectors $(\mathbf{e}_x, \mathbf{e}_y, \mathbf{e}_z)_i$ given in equation (4.27). The following transformation at $\mathbf{O}_{i+1}$ finally provides the desired unit vectors $(\mathbf{e}_{\bar{x}}, \mathbf{e}_{\bar{y}}, \mathbf{e}_{\bar{z}})_{i+1}$ of the $[\bar{x}, \bar{y}, \bar{z}]_{i+1}$ system on the reflected or refracted side at $\mathbf{O}_{i+1}$.

$$\begin{bmatrix} \mathbf{e}_{\bar{x}} \\ \mathbf{e}_{\bar{y}} \\ \mathbf{e}_{\bar{z}} \end{bmatrix}_{i+1} = \begin{bmatrix} \cos \alpha & 0 & -\sin \alpha \\ 0 & 1 & 0 \\ \sin \alpha & 0 & \cos \alpha \end{bmatrix} \begin{bmatrix} \mathbf{e}_{x_I} \\ \mathbf{e}_{y_I} \\ \mathbf{e}_{z_I} \end{bmatrix}_{i+1} , \tag{4.28}$$

with $\alpha = \pi + \varepsilon_{I,i+1} - \varepsilon_{R,i+1}$ in case of reflection and $\alpha = \varepsilon_{I,i+1} - \varepsilon_{T,i+1}$ in case of refraction.

These six steps are necessary to trace the ray from $\mathbf{O}_i$ to $\mathbf{O}_{i+1}$ and to recover the same type parameters at $\mathbf{O}_{i+1}$ as were previously computed at $\mathbf{O}_i$. One can then proceed with the tracing of the ray from $\mathbf{O}_{i+1}$ to $\mathbf{O}_{i+2}$, etc. The approach

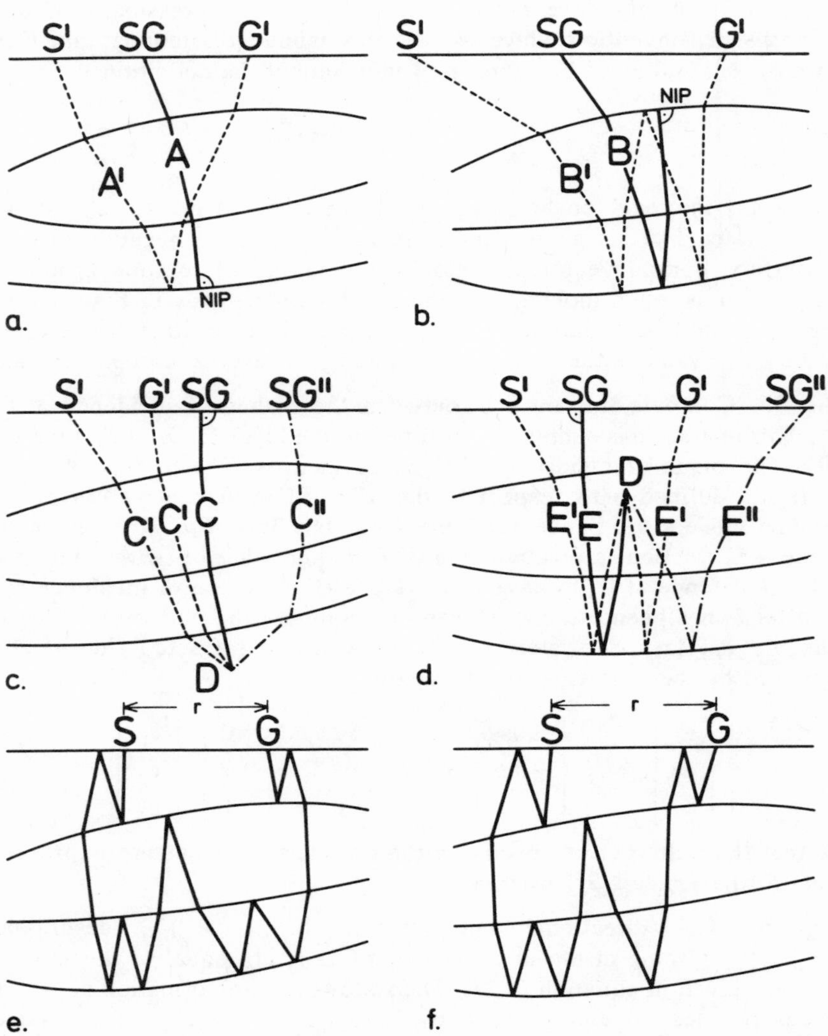

FIG. 4-5. Significant rays of the seismic reflection method featured in 2-D curved isovelocity-layer models. (a) Normal ray A; primary reflected ray A'. (b) Multiple normal ray B; multiple reflected ray B'. (c) Image ray C; finite-offset scatter ray C'; zero-offset scatter ray C''. (d) Multiple image ray E; multiple finite-offset scatter ray E'; multiple zero-offset scatter ray E''. (e) Symmetric multiple ray. (f) Asymmetric multiple ray.

described here contains only the most essential steps. These may, for instance, be extended to include *bicubic spline functions* (Ahlberg et al, 1967; Späth, 1973) for the approximation of the first-order velocity interfaces. Spline functions play an important role in the ray method.

The above algorithm can easily be extended to include layer velocities that vary linearly in accordance with the law $v(x_0, y_0, z_0) = k_x x_0 + k_y y_0 + k_z z_0$. Rays then have circular trajectories within each layer (Michaels, 1977; Hubral, 1979c). The sequence of operations described above could have been simplified for less complex subsurface models (Dürbaum, 1953; Sorrels et al, 1971; Gangi and Yang, 1976). On the other hand, it might have to be (considerably) more complicated for more complex models such as those involving arbitrary inhomogeneous or anisotropic layer media.

"Ray-tracing systems" for such models consist of coupled ordinary differential equations that determine the trajectories of rays (Kline and Kay, 1965; Vlaar, 1968; Yacoub et al, 1970; Gassmann, 1964, 1972; Psencik, 1972; Cerveny et al, 1977; Julian and Gubbins, 1977). Those more complicated systems are of particular importance for the simulation of diving waves in refraction seismology and deep-seismic sounding. Trajectories of diving waves may deviate largely from vertical, the predominant direction of wave propagation in reflection seismology.

Shah (1973a) describes a similar algorithm for efficient construction of rays. That algorithm, however, lacks some benefits gained in solving 3-D ray-tracing problems with the aid of local coordinate systems. The advantages of using these coordinate systems should become obvious later on. More is said about the moving system in section 4.3 and Appendix A.

4.2.1 Significant rays in reflection seismology

Among the multitude of rays that can be traced through subsurface velocity models, certain types of rays are particularly significant for the seismic reflection method. Although these rays are equally important in more complex isotropic media, we illustrate them here in 2-D isovelocity-layer models (Figure 4-5).

Ray A' is a *primary reflected ray* between a source S' and a separate receiver G'. It becomes the familiar (two-way path) *normal ray*, ray A, when source and receiver coincide. A normal ray is *perpendicular to the interface* to which it is traced. It can be associated with the *minimum traveltime* from the coincident source-receiver pair SG (Figure 4-5a) to the selected interface. The two-way time along a normal ray is variously known as the zero-offset (reflection) time, the t_0-time, or the two-way normal (reflection) time. Rays B' and B (Figure 4-5b) are the counterparts of rays A' and A in the case of multiple reflection.

Seismic energy is returned not only by reflection but also by wave scattering. Let point D represent a subsurface scatterer. If a source is at S' and a receiver is at G' (Figure 4-5c), the minimum traveltime for the downgoing and back-scattered wave is consumed along the paths $S'D$ and DG' (ray C'). These paths coincide if the source and receiver coincide. Ray C'' is a (zero-offset, two-way path) scatter ray for the coincident source-receiver pair SG'' (Figure 4-5c) and the subsurface scatterer D.

One particular scatter ray has great significance in migration. It is the (two-way path) image ray, ray C in Figure 4-5c. Image rays are vertical at the earth's surface (we assume here that the earth's surface is horizontal). They can be associated with the *minimum traveltime* or stationary traveltime (one-way path) from D to the earth's surface. The two-way time along an image ray is called the *two-way image time*. Image rays play a role in the theory of seismic imaging or time migration that is similar to the fundamental role played by normal rays in the theory of CDP stacking. Provided the receiver of a coincident source-receiver pair can only respond to vertically emerging waves, the image ray describes the subsurface locations from which (mostly scattered) energy is received. Most ray theoretical considerations in later chapters will involve either normal or image rays. The obvious dualism between the two will be elaborated later. Rays E, E', and E'' (Figure 4-5d) are the counterparts of rays C, C', and C'' when the raypaths of a scattered wave include a reflection.

It is useful in layered media to distinguish between *symmetric-* and *asymmetric-*multiple rays. Rays for multiple reflections are called symmetric (Figure 4-5e) if the considered downgoing and upgoing elementary waves reflect and refract at identical interfaces. Multiple rays which are not symmetric in this sense are called asymmetric (Figure 4-5f).

It is an easy matter to trace any of the ray types depicted in Figure 4-5 as long as rays do not have to go through specified end points. At any particular point in the medium, one merely starts the ray in a chosen direction and uses the six steps above to trace its path. Tracing a ray that passes through a selected location is less straightforward. In section 4.4, we shall describe an iterative approximation procedure for determining the rays that connect specified pairs of separated sources and receivers. In that particular procedure, which makes use of wavefront curvatures, a ray is sent out from the source in a trial direction. The location and direction of its emergence at the surface of the medium are used to guide the algorithm in selecting the direction of a ray to send back from the receiver toward the source. The flip-flop iteration of sending trial rays between source and receiver will generally converge toward giving the ray that connects the two to any desired accuracy.

Special rays may not require the iterative search procedure of section 4.4. Consider, for example, normal rays from a specified reflector. So long as these rays are not required to go through specific surface locations, they are easily constructed in one pass by tracing upward from the reflector. Simply, the shape of the reflector dictates the starting directions of all normal rays.

Now, if reflecting subsurface velocity boundaries are planar, a normal ray to the interface through any specified surface location can be constructed in a straightforward manner. All that is required is to trace a normal ray from an arbitrary point on the reflecting interface up to the earth's surface. Tracing a *parallel* ray down from the specified surface position then provides the desired ray.

Image rays are much easier to construct than normal rays. In fact, all such rays to a particular subsurface interface can be constructed simply in one pass. Since image rays depart vertically downward from the earth's surface, rays to any deeper interface include those to all shallower ones.

As with normal rays, if subsurface velocity interfaces are planar, an image ray through a specified subsurface point can also be constructed in two steps. An image ray is first traced from any arbitrary point on the earth's surface down into the medium. The desired image ray is then the parallel ray traced upward from the subsurface point to the surface.

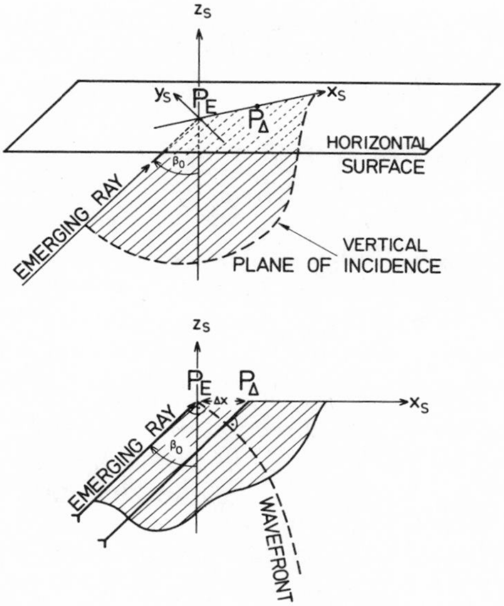

FIG. 4-6. Ray emerging at the surface of the earth. Top: isometric projection. Bottom: the plane of incidence ($y_s = 0$) is in the plane of the figure.

It is obvious that waves traveling in opposite directions along any specified ray require equal traveltimes. Though apparently trivial, this principle of reciprocity nevertheless is fundamental to several important arguments by analogy used later. Two of these analogies are described in the following. First, assume that the earth's surface is densely covered with coincident source-receiver pairs.

(1) All two-way path, normal rays from coinciding source-receiver pairs at the surface to a selected reflector can be associated with the rays of a hypothetical wavefront that originates at that reflector. The two-way normal reflection times are thus double the times consumed by a hypothetical wavefront that originates on the reflector and arrives at the surface.

(2) A signal scattered back from a subsurface scatterer D to a coincident source-receiver pair SG'' (ray C'' in Figure 4-5c) travels upward along the same scatter ray along which the wave travels down to the scattering point. The traveltime of the scattered wave recorded at SG'' thus requires twice the time of a hypothetical wave that originates at the subsurface point D.

All rays that connect surface points with the subsurface scatterer D in Figure 4-5c and Figure 4-5d can consequently be associated with a hypothetical wavefront that originates at D.

4.2.2 Traveltime gradient

The following formula [equation (4.29)] is the simplest of all the formulas that relate time measurements at the earth's surface with subsurface raypath parameters. It provides a connection between the direction of an emerging ray and the gradient of the observed *traveltime function* at the surface of an isotropic earth (Figure 4-6):

$$\nabla t(x_s, y_s)\Big|_{P_E} = \left(\frac{\partial t}{\partial x_s}, \frac{\partial t}{\partial y_s}\right)\Big|_{P_E} = \left(\frac{\sin \beta_0}{v}, 0\right). \tag{4.29}$$

Here $t(x_s, y_s)$ is the (round trip) traveltime of the emerging wavefront with respect to a Cartesian $[x_s, y_s]$ system at the ray emergence point P_E. β_0 is the *ray emergence angle*. The velocity v is the velocity at P_E in the first layer. It may be a function of space coordinates.

Following usual practice, the $[x_s, y_s]$ system is selected in such a way that the positive x_s-axis points in the direction of steepest ascent of the traveltime function. This choice involves that β_0 is always nonnegative and satisfies the condition $\beta_0 \leq \pi/2$. The system is defined entirely by either the traveltime function or the emerging ray direction since the x_s-direction agrees also with the direction of the projection of the vector of the emerging ray onto the surface. This accounts for the gradient component in the y_s-direction being zero. The equation $\partial t/\partial x_s = \sin \beta_0 / v$ is easily verified as follows.

Figure 4-6 shows the plane of incidence at $P_E(x_s = 0)$. Point P_Δ is placed near P_E at $x_s = \Delta x$. The traveltimes of an arbitrary emerging wave to the two points are given by $t(0,0)$ and $t(\Delta x, 0)$. Since wavefronts in isotropic media are perpendicular to rays, one can approximate them at P_E by their tangential plane. Within the small incident triangle formed by this tangential plane, the line $P_E P_\Delta$ and the ray arriving at P_Δ

$$v\,[t(\Delta x, 0) - t(0,0)] \approx \Delta x \sin \beta_0 .$$

In the limit where P_Δ moves toward P_E along the x_s-axis, we have

$$\frac{\partial t}{\partial x_s} = \frac{\sin \beta_0}{v}.$$

Equation (4.29) pertains to traveltimes of reflected waves from a single source. A slightly different case of particular interest in exploration seismology involves normally reflected waves from many coincident source-receiver pairs. These can be simulated by a hypothetical wavefront that originates in an interface and moves up to the surface at a velocity that is half the given local velocity. If $T(x_s, y_s)$ now describes the two-way normal time, then equation (4.29) is replaced by

$$\nabla T(x_s, y_s)\mid_{P_E} = (2 \sin \beta_0 / v, 0), \tag{4.30}$$

where the direction of a normal ray is to be determined at P_E.

According to Snell's law, the emerging ray location and direction determine the entire raypath through the subsurface. Thus, provided the velocity model and the sequence of reflecting interfaces are known, the time gradient together with the traveltime thus completely specify the entire ray from the emergent point on the surface back to the ray origin. Formula (4.29), also known as Tuchel's formula (Tuchel, 1943), will be rederived in section 4.3.6 in a more general context.

4.2.3 Summary

In section 4.2, we have done little more than provide a recursive method for tracing a ray from any selected point in some chosen direction through a 3-D isovelocity-layer model with smoothly curved first-order interfaces. We made use of a moving coordinate system that accompanies the wavefront along the selected ray. At reflecting or refracting interface points, this moving frame changes orientation in a discontinuous but well-defined manner.

We also pointed out certain types of rays that are particularly significant for the seismic reflection method. Most important for later purposes are the normal ray and image ray. We emphasized that most rays in forward ray-modeling problems must be constructed in an iterative fashion if they are to pass through specified points. Interestingly, as we shall see, a more direct approach is available for solving inverse traveltime problems.

Finally, we presented a simple formula which relates the direction of an emerging ray to the gradient of the traveltime at the ray emergence point. This relationship turns out to be a key component for solving inverse traveltime problems and guiding an emerging (real or hypothetical) wavefront backward to its source. Though it has long been a key formula for solving classical time-to-depth migration problems, formula (4.29) is equally important for computing interval velocities from traveltime measurements. As we shall see, both problems are strongly interrelated.

4.3 Wavefront curvatures

Having looked briefly at rays, let us now look at wavefronts in isovelocity layers separated by 3-D curved boundaries. We shall show that the wavefront curvature at any wavefront point can be expressed analytically in terms of seismic parameters along the ray that connects the observation point with the ray origin. We shall present three laws required for computation of wavefront curvatures anywhere along an arbitrary ray. The ray may be associated with a wavefront traveling with either P- or S-wave velocity.

The curvature laws are later used primarily to establish "Dix-type" formulas for computation of interval velocity, using a method similar to the one described by Shah (1973b) for 2-D plane dipping isovelocity layers. The most general Dix-type formula discussed here will be the one for the 3-D curved layered model of Figure 2-1. We will use curvature laws for computing interval velocities from either stacking velocities or from migration velocities and for recovering curvatures of subsurface reflectors when performing time-to-depth migration. The reader may soon appreciate the utility of these laws in facilitating the solution of various inverse seismic traveltime problems.

Curvature laws can also be used to provide approximate forward solutions for the amplitude behavior of body waves and head waves by the ray method (Cerveny and Ravindra, 1971). This method is concerned largely with the computation of *geometrical spreading*, a concept closely related to wavefront curvature; the subject is discussed briefly in the next section. Curvature laws in various forms are well known for 2- and 3-D velocity models with curved first-order interfaces (Gullstrand, 1915; Gel' chinskiy, 1961; Deschamps, 1972; Stavroudis, 1972, 1976). Their derivation in 2 × 2 matrix form with respect to our moving coordinate frame and the 3-D isovelocity layer model is given in Appendix B. Our proof is based on ray-theoretical concepts used by Krey (1976). A generalization of the laws to inhomogeneous velocity media with first-order interfaces is given by Hubral (1979a).

In the previous section, we introduced the moving $[x, y, z]$ frame. The reader should be familiar with its definition in order to follow the subsequent discussion. Let us define a further auxiliary right-hand $[x_F, y_F, z_F]$ system at all interface points of refraction $\mathbf{O}_T$ and reflection $\mathbf{O}_R$. This new system enables us to describe the interface at these points in a way suited for the derivation of the wavefront curvature laws. The plane $z_F = 0$ is tangent to the interface at $\mathbf{O}_T$ or $\mathbf{O}_R$. The positive z_F-axis coincides with the direction of the interface normal vector pointing away from the arriving wave. As a further condition, we have $y_F = y_I = \bar{y}_R = \bar{y}_T$. Thus, the $[x_F, y_F, z_F]$ system is uniquely defined at all points where a ray intersects an interface.

Let us now approximate the wavefront at **P** (Figure 4-3) by a quadratic surface in the moving frame:

$$2z = -\mathbf{XAX}^T, \tag{4.31}$$

where

$$\mathbf{X} = (x, y); \ \mathbf{A} = \begin{bmatrix} a_{11} & a_{12} \\ a_{12} & a_{22} \end{bmatrix}.$$

The choice of the minus sign in equation (4.31) will be explained later.

A is the *wavefront curvature matrix*. In particular, $\mathbf{A}_I$, $\mathbf{A}_T$, and $\mathbf{A}_R$ designate the wavefront curvature matrices of the incident, refracted, and reflected wavefronts at an interface point with respect to the moving system at that point.

$\bar{\mathbf{A}}_T$ and $\bar{\mathbf{A}}_R$ pertain to the refracted and reflected wave within the auxiliary $[\bar{x}_T, \bar{y}_T, \bar{z}_T]$ and $[\bar{x}_R, \bar{y}_R, \bar{z}_R]$ systems, respectively. Just as we use equation (4.31) to express the wavefront shape with respect to the moving $[x, y, z]$ system, we can approximate each interface at a ray intersection point with respect to the $[x_F, y_F, z_F]$ system by the following second-order polynomial:

$$2\, z_F = \mathbf{X}_F \mathbf{B} \mathbf{X}_F^T, \tag{4.32}$$

where

$$\mathbf{X}_F = (x_F, y_F); \quad \mathbf{B} = \begin{bmatrix} b_{11} & b_{12} \\ b_{12} & b_{22} \end{bmatrix}.$$

B is the interface curvature matrix. If the interface is planar, then

$$\mathbf{B} = \begin{bmatrix} 0 & 0 \\ 0 & 0 \end{bmatrix}$$

is the 2 × 2 null matrix subsequently designated as **N**. In section 4.2, as a by-product of the 3-D ray-tracing scheme, we obtained the following second-order approximation for the interface at $\mathbf{O}_{i+1}$ with respect to the $[\bar{x}, \bar{y}, \bar{z}]$ system at $\mathbf{O}_i$ [see equation (4.22)]:

$$\Phi(\bar{x}, \bar{y}, \bar{z}) = 0 = \bar{z} - \bar{\mathbf{X}}\,\bar{\mathbf{B}}\,\bar{\mathbf{X}}^T - \bar{\mathbf{C}}\,\bar{\mathbf{X}}^T - s_{i+1}. \tag{4.33}$$

In Appendix A, we show how the interface curvature matrix **B**, required for equation (4.32) at point $\mathbf{O}_{i+1}$, is obtained from equation (4.33).

4.3.1 Radius of curvature

Two kinds of curvature matrices **A** and **B** were introduced above. **A** is the wavefront curvature matrix with respect to the moving $[x, y, z]$ system. **B** is the interface curvature matrix with respect to the fixed $[x_F, y_F, z_F]$ system attached to some interface point.

Just as the $[x, y, z]$ system can be defined at every point **P** along a ray, the ensemble of planes $y = x \tan\phi$ $(0 \le \phi \le 2\pi)$ accompanying the moving coordinate system is uniquely defined as well. Let $K_A(\phi)$ be the curvature of the wavefront at **P** within the plane $y = x \tan\phi$. Then $K_A(\phi)$ can be obtained from **A** with the help of Euler's equation (Hubral, 1976b) as:

$$K_A(\phi) \equiv 1/R_A(\phi) = \mathbf{e}\,\mathbf{A}\,\mathbf{e}^T, \tag{4.34}$$

where

$$\mathbf{e} = (\cos\phi, \sin\phi),$$

$R_A(\phi)$ being the radius of wavefront curvature in the specified plane. Replacing **A** in equation (4.34) by **B** correspondingly provides the interface curvature within a plane $y_F = x_F \tan\phi$ in the $[x_F, y_F, z_F]$ system.

The inverse matrices $\mathbf{R}_A = \mathbf{A}^{-1}$ and $\mathbf{R}_B = \mathbf{B}^{-1}$, subsequently referred to as *radius matrices*, are as useful as the curvature matrices **A** and **B**. When transformed into their principal coordinate systems, radius matrices take on the diagonal form

$$\begin{bmatrix} R_x' & 0 \\ 0 & R_y' \end{bmatrix}, \tag{4.35}$$

where R_x' and R_y' are the principal radii of curvature for either the wavefront or interface. In the same principal coordinate systems, the curvature matrices also take on a diagonal form. They include then the two principal wavefront curvatures $1/R_x'$ and $1/R_y'$.

Waves propagating through curved velocity layers can have radii of wavefront curvature that become negative as well as positive. The choice of sign is a matter of definition. When a wavefront lags behind its tangential x-y-plane at **P**, then we shall define both principal radii of $\mathbf{R}_A$ as positive. If the wavefront lies ahead of the tangential plane, then both principal radii are negative. This convention explains the choice of the minus sign in equation (4.31). Likewise, both principal interface radii of $\mathbf{R}_B$ are positive if the interface appears convex to the arriving wavefront at $\mathbf{O}_T$ or $\mathbf{O}_R$. They are both negative if the interface appears concave. This explains the choice of the plus sign in equation (4.32).

The concept of the "directional sphere" was introduced by Krey (1976) to aid in transferring data related to a ray from one interface to the next. The directional sphere can be used to derive relationships between $\mathbf{A}_T$ and $\mathbf{A}_I$ as well as between $\mathbf{A}_R$ and $\mathbf{A}_I$. The law connecting $\mathbf{A}_T$ with $\mathbf{A}_I$ will be referred to as the *refraction law of curvature* or simply the refraction law. The law connecting $\mathbf{A}_R$ with $\mathbf{A}_I$ is correspondingly called the *reflection law (of curvature).* Both laws include the interface curvature matrix **B**. Apart from expressions that describe the change of wavefront curvature at an interface, one must know as well how curvatures change during propagation through a homogeneous layer. This change is described by the *transmission law (of curvature).*

4.3.2 Transmission law of curvature

The transmission law is the simplest of the curvature laws. It can be stated as:

$$\mathbf{R}_{P_2} = \mathbf{R}_{P_1} + v\,\Delta t\,\mathbf{I}. \tag{4.36}$$

Here, $\mathbf{R}_{P_1}$ is the radius matrix at ray point $\mathbf{P}_1$ and $\mathbf{R}_{P_2}$ at ray point $\mathbf{P}_2$ with respect to the moving frame; $\mathbf{P}_1$ and $\mathbf{P}_2$ are connected by a straight raypath segment in a layer having constant velocity v; Δt is the time required for the wave to travel from $\mathbf{P}_1$ to $\mathbf{P}_2$. **I** is the 2×2 unit matrix. If transformed to principal axes, equation (4.36) reduces to the following more obvious relationship for each principal radius of curvature:

$$R'_{P_2} = R'_{P_1} + v\,\Delta t. \tag{4.37}$$

R'_{P_1} and R'_{P_2} are principal radii with respect to parallel principal axes at $\mathbf{P}_1$ and $\mathbf{P}_2$. Unlike the principal radii of curvature, other radii of curvature along a straight ray need not increase linearly with distance (see Appendix E). It can be shown that equation (4.36) results from a Riccati-type differential equation that describes the change of wavefront curvature in a general inhomogeneous medium (Hubral, 1979b).

Principal coordinate systems need never be explicitly constructed to solve any problem discussed in this work. This simplification arises from our use of the well-defined moving $[x, y, z]$ frame, which will generally differ from the principal coordinate system that could be associated with the moving wavefront.

4.3.3 Refraction law of curvature

At a refracting interface point $\mathbf{O}_T$, the matrices $\mathbf{A}_I$, $\mathbf{A}_T$, and **B** are defined with respect to the $[x_I, y_I, z_I]$, $[x_T, y_T, z_T]$, and $[x_F, y_F, z_F]$ systems, whereby the first two systems are part of the moving frame. One can show (Appendix B) that $\mathbf{A}_I$, $\mathbf{A}_T$, and **B** are connected with each other through the rather complex appearing expression:

$$\mathbf{A}_T = \mathbf{D}^{-1}\left(\frac{v_T}{v_I}\,\mathbf{S}\,\mathbf{A}_I\,\mathbf{S} + \rho\,\mathbf{S}_T^{-1}\,\mathbf{B}\,\mathbf{S}_T^{-1}\right)\mathbf{D}, \tag{4.38}$$

where

$$\mathbf{D} = \begin{bmatrix} \cos\delta & -\sin\delta \\ \sin\delta & \cos\delta \end{bmatrix},$$

$$\mathbf{S} = \begin{bmatrix} \cos \varepsilon_I / \cos \varepsilon_T & 0 \\ 0 & 1 \end{bmatrix},$$

$$\mathbf{S}_T = \begin{bmatrix} \cos \varepsilon_T & 0 \\ 0 & 1 \end{bmatrix},$$

and

$$\rho = \frac{v_T}{v_I} \cos \varepsilon_I - \cos \varepsilon_T.$$

Note that equation (4.38) is independent of the signs of ε_I and ε_T, required for the proper construction of the moving frame.

Matrices $\mathbf{A}_T$ and $\mathbf{A}_I$ pertain to the $[x, y, z]$ systems on the two sides of $\mathbf{O}_T$. v_I and ε_I refer to the incident side and v_T and ε_T to the refracted side. The matrix $\mathbf{D}$ occurs because $\mathbf{A}_T$ pertains to the $[x_T, y_T, z_T]$ system rather than the $[\bar{x}_T, \bar{y}_T, \bar{z}_T]$ system.

The first term in equation (4.38) describes the wavefront curvature change that would result if the interface were a plane ($\mathbf{B} = \mathbf{N}$)—the tangential plane to the interface at the intersection point. The second term provides the change in wavefront curvature due to interface curvature alone. It can be interpreted as the contribution to the curvature of the refracted wavefront resulting when a plane incident wave ($\mathbf{A}_I = \mathbf{N}$) arrives at $\mathbf{O}_T$.

4.3.4 Reflection law of curvature

At a reflecting interface point $\mathbf{O}_R$, the matrices $\mathbf{A}_I$, $\mathbf{A}_R$, and $\mathbf{B}$ are defined with respect to the $[x_I, y_I, z_I]$, $[x_R, y_R, z_R]$, and $[x_F, y_F, z_F]$ systems, respectively. In Appendix B we show that the three curvature matrices are then connected with each other as follows:

$$\mathbf{A}_R = \mathbf{D}^{-1} \mathbf{I}_R \left(\frac{v_R}{v_I} \mathbf{S}' \mathbf{A}_I \mathbf{S}' + \rho' \mathbf{S}_R^{-1} \mathbf{B} \mathbf{S}_R^{-1} \right) \mathbf{I}_R \mathbf{D}, \qquad (4.39)$$

$$\mathbf{I}_R = \begin{bmatrix} -1 & 0 \\ 0 & 1 \end{bmatrix},$$

$$\mathbf{S}' = \begin{bmatrix} -\cos \varepsilon_I / \cos \varepsilon_R & 0 \\ 0 & 1 \end{bmatrix},$$

$$\mathbf{S}_R = \begin{bmatrix} -\cos \varepsilon_R & 0 \\ 0 & 1 \end{bmatrix},$$

$$\rho' = \frac{v_R}{v_I} \cos \varepsilon_I + \cos \varepsilon_R.$$

Note again that the signs for the angles ε_I and ε_R do not affect the result.

The reflection law (4.39) is a special case of the refraction law (4.38); it can be obtained directly from (4.38) by replacing $\mathbf{A}_T$ with $\mathbf{A}_R$, v_T with v_R, and ε_T with $\varepsilon_R - \pi$. The contribution due to interface curvature can again be viewed as being that for a plane wavefront ($\mathbf{A}_I = \mathbf{N}$) that arrives at $\mathbf{O}_R$.

The matrix **D** in equation (4.39) has the same meaning as in equation (4.38). The matrix $\mathbf{I}_R$ in equation (4.39) accounts for the "mirror inversion" caused by a reflecting interface. The reflection law becomes simpler if no mode-converted waves are considered; then $v_R = v_I$, and $\varepsilon_R = -\varepsilon_I$.

Plane reflecting interfaces act like plane mirrors and cause no distortion of a nonconverted reflected wave. When all interfaces are planar, one can employ principles of *plane-interface imaging* (see section 4.5.3) to simplify various considerations. For example, plane-interface imaging makes the reflection law for a plane interface (of arbitrary dip and strike) redundant for a nonconverted, reflected wave.

4.3.5 Wavefront curvature along a raypath from a point source

The three matrix laws presented above are sufficient for the computation of the wavefront curvature at any point along a ray traced through a 3-D layered earth model. Starting from a source with a known wavefront curvature matrix, we can apply the three laws in cascade to find an analytic expression for the curvature matrix of the wavefront at any other desired ray point. If a ray originates at a radially symmetric *point source,* then the radius matrix at the ray origin is $\mathbf{R} = \mathbf{A}^{-1} = \mathbf{N}$. If, on the other hand, it originates as a plane wave, then $\mathbf{R}^{-1} = \mathbf{A} = \mathbf{N}$. We should note that if the radius of curvature of a structure becomes less than the radius of curvature of the impinging wavefront, marked diffractions appear. This situation can easily be flagged by the recursion described below.

The recursive application of the matrix laws for a selected ray provides a resulting wavefront curvature matrix that is always expressed in the $[x, y, z]$ system at each point along the ray. The appropriate use of curvature laws hence requires appropriate construction of the moving frame. Provided that both the $[x, y, z]$ system and the matrix **A** are known at some ray point, then $2z = -\mathbf{XAX}^T$ provides the second-order approximation to the wavefront there.

Figure 4-3 shows a ray from a point source at S to a receiver at G. It is assumed that this ray has already been traced. To obtain the curvature matrix of the emerging wavefront at G, one can follow the recursion procedure described below. Only the first few steps of the recursion are formulated as they are sufficient to demonstrate the method. The following notation is used: (ε_I, ε_T, and ε_R from (4.38) and (4.39) are replaced by α_i and β_i)

$$\mathbf{D}_i = \begin{bmatrix} \cos\delta_i & -\sin\delta_i \\ \sin\delta_i & \cos\delta_i \end{bmatrix},$$

$$\mathbf{S}_i = \begin{bmatrix} \cos\alpha_i/\cos\beta_i & 0 \\ 0 & 1 \end{bmatrix},$$

$$\mathbf{S}'_i = \begin{bmatrix} -\cos\alpha_i/\cos\beta_i & 0 \\ 0 & 1 \end{bmatrix},$$

$$\mathbf{S}_{T,i} = \begin{bmatrix} \cos\beta_i & 0 \\ 0 & 1 \end{bmatrix},$$

$$\mathbf{S}_{R,i} = \begin{bmatrix} -\cos\beta_i & 0 \\ 0 & 1 \end{bmatrix},$$

$$\rho_i = \frac{v_{i+1}}{v_i}\cos\alpha_i - \cos\beta_i,$$

and

$$\rho_i' = \frac{v_{i+1}}{v_i}\cos\alpha_i + \cos\beta_i.$$

Index i pertains to the ith interface. Ray segments and interfaces are counted successively from S and G, starting with $i = 1$; $\mathbf{B}_i$ is the interface curvature matrix at point $\mathbf{O}_i$, the end point of the ith segment; s_i is the length of the raypath segment from $\mathbf{O}_{i-1}$ to $\mathbf{O}_i$; α_i is the incidence angle of the ith raypath segment at $\mathbf{O}_i$; β_i is the corresponding refraction or reflection angle; and v_i is the interval velocity along the ith ray segment. The moving $[x, y, z]$-coordinate frame at the source point S is chosen in such a way that the x-axis falls in the plane of incidence at $\mathbf{O}_1$. Wavefront curvature matrices with respect to the moving frame can now be computed as follows:

(1) Curvature matrix on the incident side of $\mathbf{O}_1$, according to equations (4.36) and (4.37)

$$\mathbf{A}_{I,1} = (1/s_1)\,\mathbf{I}. \tag{4.40a}$$

(2) Curvature matrix on refracted side of $\mathbf{O}_1$, according to equation (4.38)

$$\mathbf{A}_{T,1} = \mathbf{D}_1^{-1}\,[(v_2/v_1)\,\mathbf{S}_1\,\mathbf{A}_{I,1}\,\mathbf{S}_1 + \rho_1\,\mathbf{S}_{T,1}^{-1}\,\mathbf{B}_1\,\mathbf{S}_{T,1}^{-1}]\,\mathbf{D}_1. \tag{4.40b}$$

(3) Curvature matrix on incident side of $\mathbf{O}_2$, according to equation (4.36)

$$\mathbf{A}_{I,2}^{-1} = \mathbf{A}_{T,1}^{-1} + s_2\,\mathbf{I}. \tag{4.40c}$$

(4) Curvature matrix on refracted side of $\mathbf{O}_2$, according to equation (4.38)

$$\mathbf{A}_{T,2} = \mathbf{D}_2^{-1}\,[(v_3/v_2)\,\mathbf{S}_2\,\mathbf{A}_{I,2}\,\mathbf{S}_2 + \rho_2\,\mathbf{S}_{T,2}^{-1}\,\mathbf{B}_2\,\mathbf{S}_{T,2}^{-1}]\,\mathbf{D}_2. \tag{4.40d}$$

(5) Curvature matrix on incident side of $\mathbf{O}_3$, again according to equation (4.36)

$$\mathbf{A}_{I,3}^{-1} = \mathbf{A}_{T,2}^{-1} + s_3\,\mathbf{I}. \tag{4.40e}$$

(6) Curvature matrix on reflected side of $\mathbf{O}_3$, according to equation (4.39)

$$\mathbf{A}_{R,3} = \mathbf{D}_3^{-1}\,\mathbf{I}_R\,[(v_4/v_3)\,\mathbf{S}_3'\,\mathbf{A}_{I,3}\,\mathbf{S}_3' + \rho_3'\,\mathbf{S}_{R,3}^{-1}\,\mathbf{B}_3\,\mathbf{S}_{R,3}^{-1}]\,\mathbf{I}_R\,\mathbf{D}_3. \tag{4.40f}$$

This recursive procedure is continued in an obvious manner in order eventually to find the curvature matrix at point G with respect to the emerging moving frame. It is helpful to view the recursion in conjunction with the moving wavefront. A forward recursion will then correspond to a forward propagating (expanding) wavefront and a backward recursion to a backward propagating (shrinking) wavefront.

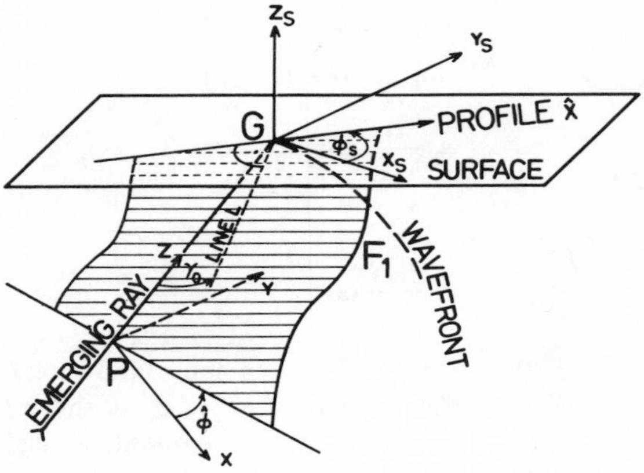

FIG. 4-7. The emerging ray and profile define plane F_1. Plane F_1 would be vertical only if the profile direction $\hat{x}$ coincided with the direction x_s.

4.3.6 Emerging wavefront

Figure 4–7 shows a ray that emerges at the surface of the earth at point G. The moving system is indicated at some point **P** on the ray within the layer nearest the surface of the earth. At the surface of the earth, a right-hand $[x_s, y_s, z_s]$-system is placed as shown previously in Figure 4–6. As shown in section 4.2.2, the $[x_s, y_s, z_s]$-system can be entirely derived from the direction of the emerging ray: z_s points in the upward direction and y_s in the direction of isochrons; i.e., y_s is parallel to y.

The second-order approximation to the shape of the emerging wavefront at G can be used to find a second-order approximation for the traveltime of the wave in the vicinity of G along a profile at any azimuth angle ϕ_s relative to the x_s-axis. The distance measured from G along the profile is denoted by the variable $\hat{x}$. The traveltime $t(\hat{x})$, recorded along the profile, pertains to the wavefront emerging within the plane F_1, defined by the profile and the emerging ray. This plane is generally not vertical (Figure 4–7).

The acute angle between the emerging ray and line L, perpendicular to the profile within F_1, is γ_0. It is defined to be positive when line L falls between the forward profile line direction $\hat{x}$ and the emerging ray direction. If γ_0 and the radius of curvature R_0 of the emerging wave within plane F_1 are known, then the small-$\hat{x}$ approximation for traveltime along the profile can be obtained from either of the following two equations:

$$t(\hat{x}) = t(0) + \frac{\sin \gamma_0}{v_1} \hat{x} + \frac{\cos^2 \gamma_0}{2 v_1 R_0} \hat{x}^2 + \ldots, \tag{4.41}$$

or

$$t^2(\hat{x}) = \left[t(0) + \frac{\sin \gamma_0}{v_1} \hat{x} \right]^2 + \frac{t(0) \cos^2 \gamma_0}{v_1 R_0} \hat{x}^2 + \ldots, \tag{4.41a}$$

where powers of $\hat{x}$ higher than two are neglected.

With the help of rays emerging in plane F_1 of Figure 4–7, one can deduce the proof for equation (4.41) [see also Shah (1973b)]. First we have,

$$\lim_{\Delta\hat{x}\to 0} \frac{v_1 \Delta t}{\Delta \hat{x}} = v_1 \frac{\partial t}{\partial \hat{x}} = \sin \gamma_0, \tag{4.42}$$

and

$$\lim_{\Delta\hat{x}\to 0} \frac{R_0 \Delta \gamma_0}{\Delta \hat{x}} = R_0 \frac{\partial \gamma_0}{\partial \hat{x}} = \cos \gamma_0. \tag{4.42a}$$

Now, differentiating equation (4.42) with respect to $\hat{x}$ and substituting $\partial\gamma_0/\partial\hat{x}$ from equation (4.42a) gives

$$\frac{\partial^2 t}{\partial \hat{x}^2} = \frac{\cos^2 \gamma_0}{v_1 R_0}. \tag{4.43}$$

R_0 is obtained from equation (4.34) as

$$1/R_0 = \hat{\mathbf{e}} \mathbf{A}_0 \hat{\mathbf{e}}^T; \ \hat{\mathbf{e}} = (\cos \hat{\phi}, \sin \hat{\phi}), \tag{4.44}$$

$\mathbf{A}_0$ being the emerging wavefront curvature matrix at G and angle $\hat{\phi}$ being the angle defining the plane F_1 in the emerging $[x,y,z]$ system; i.e., $y = x \tan \hat{\phi}$ is the equation of F_1.

Let us now make use of the angle β_0, the positive acute angle between the emerging ray and the normal to the surface, as mentioned earlier. If angles ϕ_s and β_0 are known, one can easily compute γ_0 and $\hat{\phi}$ from the following two equations:

$$\sin \gamma_0 = \sin \beta_0 \cos \phi_s, \tag{4.45}$$

and

$$\tan \phi_s = \cos \beta_0 \tan \hat{\phi}. \tag{4.46}$$

Given the direction of the emerging ray and the curvature matrix $\mathbf{A}_0$ at G, one can now obtain a second-order approximation for the traveltime in the vicinity of G. Using equations (4.41) and (4.41a), we will do so with respect to the $[x_s, y_s]$ system.

By making use of equations (4.44)–(4.46), we obtain the following equivalent expressions for equations (4.41) and (4.41a):

$$t(x_s, y_s) = t(0,0) + \frac{\sin \beta_0}{v_1} x_s + \frac{1}{2v_1} \mathbf{X}_s \mathbf{S}_0 \mathbf{A}_0 \mathbf{S}_0 \mathbf{X}_s^T + \ldots, \tag{4.47}$$

and

$$t^2(x_s, y_s) = \left[t(0,0) + \frac{\sin \beta_0}{v_1} x_s \right]^2 + \frac{t(0,0)}{v_1} \mathbf{X}_s \mathbf{S}_0 \mathbf{A}_0 \mathbf{S}_0 \mathbf{X}_s^T + \ldots, \tag{4.47a}$$

where

$$\mathbf{X}_s = (x_s, y_s) \quad ; \quad \mathbf{S}_0 = \begin{bmatrix} \cos \beta_0 & 0 \\ 0 & 1 \end{bmatrix}. \tag{4.48}$$

Let us now consider inverse problems based upon observed traveltimes at the earth's surface. Initially we do not know the orientations of the x_s- and y_s-axes. Gradients of $t(x_s, y_s)$ are required in two different directions at G in order to establish the direction of the emerging ray in the uppermost layer and construct the $[x_s, y_s, z_s]$-system. Recall that, knowing the emerging ray, we can trace the ray back to its origin (provided no reflections have occurred or the sequence of the reflecting boundaries is known). If, in addition, second-order derivatives of the observed traveltime function are known in three different directions through G, we can determine the curvature matrix $\mathbf{A}_0$ at G. With $\mathbf{A}_0$ recovered from the two first-order and three second-order derivatives of traveltime measurements, we can, by recursively applying the curvature laws, find the wavefront curvatures anywhere else along the considered ray. In other words, the curvature laws provide the tools for propagating recorded wavefronts (strictly speaking, their second-order approximations), as obtainable from traveltimes backward in space and time. As we shall see, this reversal of wave propagation provides the key to the solution of various inverse traveltime problems.

Some readers might guess that still higher-order derivatives of observed traveltimes may lead to better definition of traveltimes and wavefronts. Wavefront approximations could indeed be developed into higher-order terms as a function of parameters found along the ray. Any attempt in this direction, however, would encounter large theoretical and insurmountable practical difficulties, as already indicated in section 4.1 where we considered the much simpler, horizontally layered model.

4.3.7 Applications

The curvature matrix laws, as given above, are valid for 3-D models with constant velocity media bounded by arbitrarily curved interfaces. They pertain to the moving coordinate system and allow for the recursive computation of wavefront curvatures at any selected ray point. Wavefront curvatures are influenced by ray segment lengths; and by incidence, reflection, refraction, and rotation (δ) angles as well as by interface curvature matrices. Radius matrices and curvature matrices are inverses of one another.

The intersection of a wavefront with any plane containing a ray is called a normal section of the wavefront. There exist two perpendicular principal normal sections which are associated with the smallest and largest value of the radius of wavefront curvature. The intersections of these normal sections with one another and with the tangential plane to the wavefront define the principal coordinate system. Such principal coordinate systems are not required in order to apply recursively the various laws. If a common dip direction exists for the layering and a profile lies along this direction, all rays are confined to a common plane, the so-called vertical seismic plane. In this plane, curvature matrices have a diagonal form, and it is sufficient to operate solely with the principal radii of curvature only.

We shall now illustrate use of the above curvature formulas in connection with less complex subsurface models that are special cases of the 3-D model. The following four examples should give the reader some idea of the key role played by the curvature laws in forward ray-modeling problems.

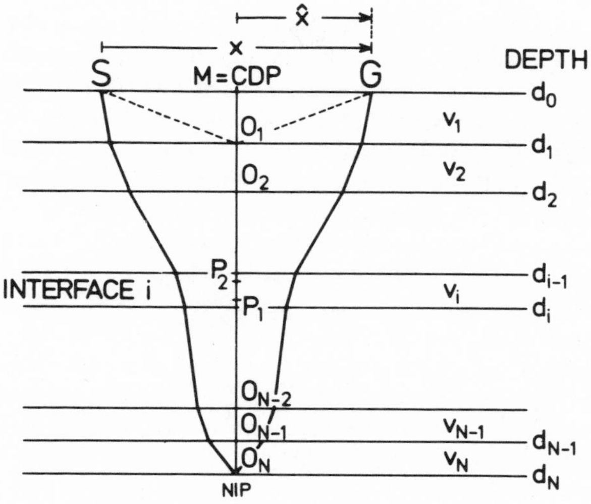

FIG. 4-8. Horizontal, isovelocity-layer model.

Example 1.—Figure 4-8 shows N horizontal isovelocity layers. A source S and a receiver G are placed symmetrically with respect to a common midpoint M and are separated by the distance x. Rays reflected at the Nth interface can be traced for different values of x. They are symmetric with respect to the vertical through M and pass through the *common reflection point* O_N. We desire a formula for approximating the reflection time $T(x)$ from S to G for small values of x; i.e., omitting the third- and higher-order terms of x. This formula is to be derived with curvature laws.

The ensemble of rays in Figure 4-8 can be associated with a hypothetical wavefront that has a point source at O_N (also labeled NIP, for normal incidence point). Let the traveltime to the surface for this wavefront be $t(\hat{x})$, where $\hat{x}$ is the distance from M to G. It is clear that $x = 2\hat{x}$, and $T(x) = 2t(\hat{x})$. We shall now compute the curvature of the wavefront of a hypothetical wave traveling upward from O_N to M. Because of symmetry about the vertical in this model, the curvature is independent of azimuth. For the horizontally layered model, the transmission law (4.36) and refraction law (4.38) reduce to the following.

Transmission law:

$$R_{P_2} = R_{P_1} + v\,\Delta t. \tag{4.49}$$

R_{P_1} and R_{P_2} are the radii of wavefront curvature at two points P_1 and P_2 on the vertical path within a layer of velocity v; Δt is the time required by the wave to travel from P_1 to P_2.

Refraction law:

$$R_T = \frac{v_I}{v_T} R_I. \tag{4.50}$$

R_T and R_I are the radii of wavefront curvature on the refracted side (velocity $= v_T$) and incident side (velocity $= v_I$) of an interface, respectively. The radius of curvature as the wavefront expands upward from NIP to M can be computed by the following simple recursion, where Δt_i denotes the one-way vertical time within the ith layer. $R_{I,j}$ and $R_{T,j}$ denote the radii of curvature on the lower and upper sides of the jth interface, respectively.

Wave on lower side of $(N-1)$th interface:

$$R_{I,N-1} = v_N \Delta t_N .$$

Wave on upper side of $(N-1)$th interface:

$$R_{T,N-1} = \frac{1}{v_{N-1}} [v_N^2 \Delta t_N] .$$

Wave on lower side of $(N-2)$th interface:

$$R_{I,N-2} = \frac{1}{v_{N-1}} [v_{N-1}^2 \Delta t_{N-1} + v_N^2 \Delta t_N] .$$

Wave on upper side of $(N-2)$th interface:

$$R_{T,N-2} = \frac{1}{v_{N-2}} [v_{N-1}^2 \Delta t_{N-1} + v_N^2 \Delta t_N] .$$

It is clear how the recursion is to be continued to obtain the radius of curvature R_0 of the hypothetical wave emerging at M:

$$R_0 = \frac{1}{v_1} (v_1^2 \Delta t_1 + v_2^2 \Delta t_2 + \ldots + v_N^2 \Delta t_N), \tag{4.51}$$

or compactly,

$$R_0 = \frac{1}{v_1} \sum_{j=1}^{N} v_j^2 \Delta t_j .$$

By using equation (4.41), one can approximate the arrival time $t(\hat{x})$ of the hypothetical wave as:

$$\begin{aligned} t(\hat{x}) &= t(0) + \frac{1}{2 v_1 R_0} \hat{x}^2 + \ldots, \\ &= t(0) + \frac{\hat{x}^2}{2 t(0)\, V_{\mathrm{RMS}}^2}, \end{aligned}$$

where

$$V_{\mathrm{RMS}}^2 = v_1 R_0 / t(0),$$

or

$$V_{\mathrm{RMS}}^2 = \frac{1}{t(0)} \sum_{j=1}^{N} v_j^2 \Delta t_j .$$

Multiplying this equation by 2, and using $x = 2\hat{x}$, we obtain the approximation for the total time from S to NIP to G,

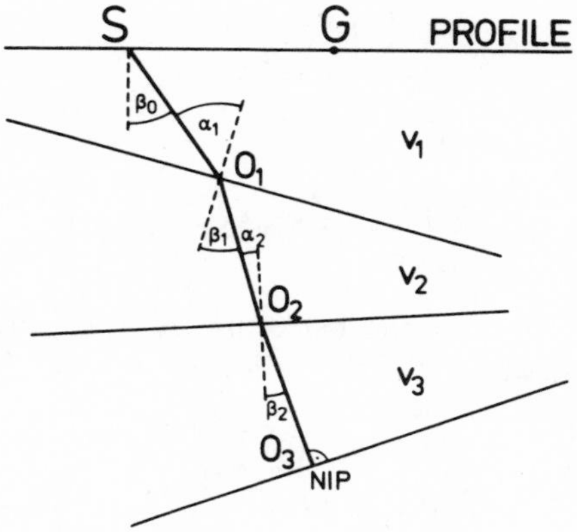

FIG. 4-9. A normal ray in a 2-D plane-dipping, isovelocity-layer model.

$$T(x) = T(0) + \frac{x^2}{2T(0)\, V_{\mathrm{RMS}}^2} + \ldots$$

or, squaring,

$$T^2(x) = T^2(0) + x^2 / V_{\mathrm{RMS}}^2 + \ldots . \tag{4.52}$$

Equation (4.52) is the well-known formula of Dix (1955). V_{RMS} is the familiar RMS velocity, and $T(0)$ is the two-way reflection time for zero offset, $x = 0$.

Example 2.—Figure 4-9 shows a 2-D model for $N = 3$ plane-dipping layers. Also featured is a normal incidence ray to the Nth reflector. A receiver G is near the source S along a profile perpendicular to the common strike direction of all interfaces. A radially symmetric wave expands from S. We first desire the radius of curvature, within the vertical seismic plane, of the reflected wave at S. It can be obtained in a recursive fashion about the normal incidence ray from S to NIP and back up to S. The transmission law is given in equation (4.49). The two additional formulas needed to solve the problem follow.

Refraction law:

$$R_T = \frac{v_I}{v_T} \frac{\cos^2 \epsilon_T}{\cos^2 \epsilon_I} R_I . \tag{4.53}$$

Reflection law:

$$R_R = R_I . \tag{4.54}$$

The radius of wavefront curvature of the primary reflected P-wave at S is obtained recursively in the following way, where Δt_i denotes the one-way time of the normal ray in the ith layer.

Wave on incident side of 0_1:

$$R_{I,1} = v_1 \Delta t_1 .$$

Wave on refracted side of 0_1:

$$R_{T,1} = \frac{1}{v_2}\left(\frac{\cos^2\beta_1}{\cos^2\alpha_1} v_1^2 \Delta t_1\right).$$

Wave on incident side at 0_2:

$$R_{I,2} = \frac{1}{v_2}\left(\frac{\cos^2\beta_1}{\cos^2\alpha_1} v_1^2 \Delta t_1 + v_2^2 \Delta t_2\right).$$

Continuing the steps downward and back upward along the normal ray provides the desired radius of curvature for the reflected wave at S:

$$\begin{aligned} R_0 = \frac{2}{v_1}\Bigg[& v_1^2\Delta t_1 + \left(\frac{\cos^2\alpha_1}{\cos^2\beta_1}\right) v_2^2 \Delta t_2 + \ldots \\ & + \left(\frac{\cos^2\alpha_1 \cos^2\alpha_2 \ldots \cos^2\alpha_{j-1}}{\cos^2\beta_1 \cos^2\beta_2 \ldots \cos^2\beta_{j-1}}\right) v_j^2 \Delta t_j \\ & + \ldots + \left(\frac{\cos^2\alpha_1 \cos^2\alpha_2 \ldots \cos^2\alpha_{N-1}}{\cos^2\beta_1 \cos^2\beta_2 \ldots \cos^2\beta_{N-1}}\right) v_N^2 \Delta t_N \Bigg], \end{aligned}$$

or, more compactly,

$$R_0 = \frac{2}{v_1} \sum_{j=1}^{N} v_j^2 \Delta t_j \prod_{k=1}^{j-1} \left(\frac{\cos^2\alpha_k}{\cos^2\beta_k}\right). \tag{4.55}$$

where

$$\prod_{k=1}^{0} = 1.$$

Rather than applying the recursion down and up along the normal ray, one can as well consider the principles of plane-interface imaging (as described in section 4.5.4), construct an image of the source and layers with respect to the Nth reflector, and perform the recursion only in one direction without considering the reflection law.

Substituting equation (4.55) into equation (4.41a) leads to Dürbaum's formula (Dürbaum, 1954)

$$t^2(x) = \left[t(0) + \frac{\sin\beta_0}{v_1} x\right]^2 + \frac{t(0)\cos^2\beta_0}{v_1 R_0} x^2 + \ldots, \tag{4.56}$$

which was originally derived in a more complex manner without the aid of wavefront curvature laws. $t(0)$ is the two-way normal time for a coincident source receiver pair at S.

It is interesting to note that for a source at NIP, the radius of curvature of the emerging wavefront at S is $R_0/2$ within the vertical plane. We established compact expression (4.55) with the help of curvature laws just for the sake of giving a simple proof of Dürbaum's formula. Such closed-form expressions become increasingly more complicated for more complex models. In fact, it is never necessary to use closed-form expressions to solve any of the problems discussed in this monograph. It is much easier to compute curvatures recursively along selected rays. One or several unknown quantities, as e.g., v_N, may well be carried through the various steps of recursion.

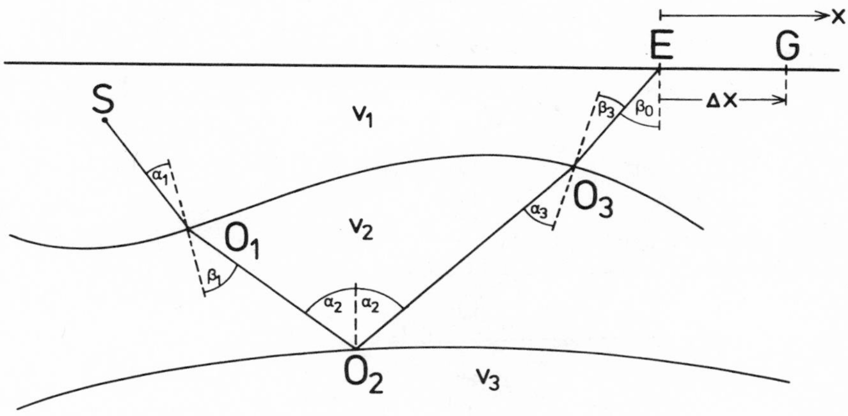

FIG. 4-10. A primary reflected ray in a 2-D isovelocity-layer model having curved interfaces.

Example 3.—Figure 4-10 shows a 2-D isovelocity-layer model with arbitrarily curved interfaces. A ray reflected at the second interface is traced from S to E. We desire an expression for the approximate traveltime from S to G, a point close to E. If the radius of curvature of the emerging wave at E is designated as R_E, then by (4.41a) a second-order traveltime approximation in the vicinity of E is available as:

$$t^2(x) = \left[t(0) + \frac{\sin \beta_0}{v_1} x \right]^2 + \frac{t(0) \cos^2 \beta_0}{v_1 R_E} x^2. \tag{4.57}$$

The x-axis has its origin at E and points in the direction of increasing traveltime. If Δx is the distance from E to G, then the traveltime from S to G is approximated by $t(\Delta x)$. To compute R_E, we require the following two formulas, special cases of equations (4.38) and (4.39).

Refraction law:

$$\frac{1}{R_T} = \frac{v_T \cos^2 \epsilon_I}{v_I \cos^2 \epsilon_T} \frac{1}{R_I} + \frac{1}{\cos^2 \epsilon_T} \left[\frac{v_T}{v_I} \cos \epsilon_I - \cos \epsilon_T \right] \frac{1}{R_F}. \tag{4.58}$$

Reflection law:

$$\frac{1}{R_R} = \frac{v_R \cos^2 \epsilon_I}{v_I \cos^2 \epsilon_R} \frac{1}{R_I} + \frac{1}{\cos^2 \epsilon_R} \left[\frac{v_R}{v_I} \cos \epsilon_I + \cos \epsilon_R \right] \frac{1}{R_F}. \tag{4.59}$$

R_F is the radius of interface curvature. It is positive if the interface appears convex to the arriving wave.

The first few steps of the recursion along the reflected ray for an unconverted wave from S to E are:

Incident wave at 0_1:

$$R_{I,1} = v_1 \Delta t_1 .$$

Refracted wave from 0_1:

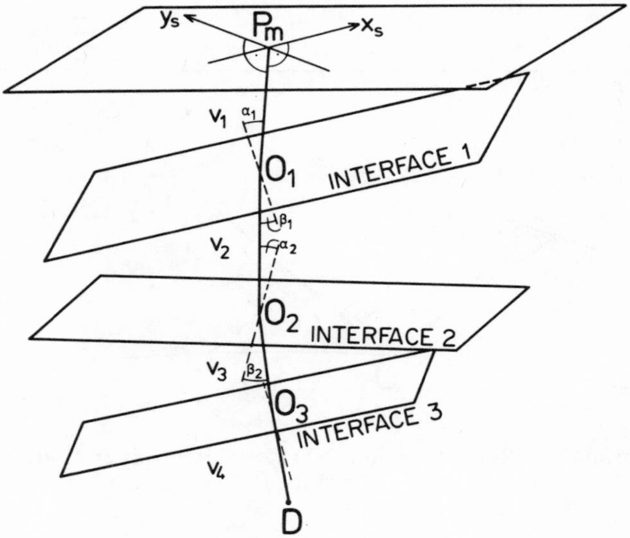

FIG. 4-11. An image ray in a 3-D plane-dipping, isovelocity-layer model.

$$\frac{1}{R_{T,1}} = \frac{v_2 \cos^2 \alpha_1}{v_1 \cos^2 \beta_1} \frac{1}{R_{I,1}} + \frac{1}{\cos^2 \beta_1} \left(\frac{v_2}{v_1} \cos\alpha_1 - \cos\beta_1 \right) \frac{1}{R_{F,1}}.$$

Incident wave at 0_2:

$$R_{I,2} = R_{T,1} + v_2 \Delta t_2.$$

Reflected wave from 0_2 ($v_R = v_I$):

$$\frac{1}{R_{R,2}} = \frac{1}{R_{I,2}} + \frac{2}{\cos \alpha_2} \frac{1}{R_{F,2}}.$$

Δt_j is the one-way traveltime along segment j of the ray from S to 0_2 to E.

Again, the recursive procedure to find R_E for use in formula (4.57) is obvious. It is possible, but unnecessary, to express R_E in closed form (a continued fraction).

Example 4.—A scattering source D (e.g., a secondary source as exemplified by a point diffractor) falls below the upper $N - 1$ plane, isovelocity layers of different dip and strike. The minimum traveltime from the source D to the surface of the earth is obtained along the image ray (Figure 4-11). At the surface, an $[x_s, y_s, z_s]$ system is placed with its origin at the image ray emergence point P_m in such a way that the x_s-axis is defined by the intersection of the incident plane at 0_1 with the surface of the earth. The second-order approximation for the arrival times of the wave with respect to the (x_s, y_s)-coordinates can be written as:

$$t^2(x_s, y_s) = t^2(0,0) + \frac{t(0,0)}{v_1} \mathbf{X}_s \mathbf{R}_0^{-1} \mathbf{X}_s^T. \tag{4.60}$$

This results from equation (4.47a), with $\beta_0 = 0$.

$\mathbf{R}_0$ is the radius matrix of the emerging wave at P_m. The two matrix curvature laws required to compute $\mathbf{R}_0$ in equation (4.60) are, according to equations (4.36) and (4.38):

Transmission law:

$$\mathbf{R}_{P_2} = \mathbf{R}_{P_1} + v\,\Delta t\,\mathbf{I}. \tag{4.61}$$

Refraction law:

$$\mathbf{R}_T = \mathbf{D}^{-1}\,[(v_I/v_T)\,\mathbf{S}^{-1}\mathbf{R}_I\mathbf{S}^{-1}]\,\mathbf{D}. \tag{4.62}$$

These are applied recursively from bottom to top along the image ray. Applying the radius matrix laws (4.61) and (4.62) recursively from D to P_m and substituting the results from one step into the next leads to the compact expression:

$$\mathbf{R}_0 = \frac{1}{v_1}\sum_{j=1}^{N} v_j^2\,\Delta t_j\,\mathbf{I}\prod_{k=1}^{j-1}(\mathbf{D}_k^{-1}\mathbf{S}_k^{-1})\prod_{k=1}^{j-1}(\mathbf{S}_{j-k}^{-1}\mathbf{D}_{j-k}), \tag{4.63}$$

where

$$\mathbf{D}_k = \begin{bmatrix} \cos\delta_k & -\sin\delta_k \\ \sin\delta_k & \cos\delta_k \end{bmatrix}; \text{ and } \mathbf{S}_k = \begin{bmatrix} \cos\beta_k/\cos\alpha_k & 0 \\ 0 & 1 \end{bmatrix}.$$

β_k is now the angle of incidence (lower side of interface k), whereas α_k is the angle of refraction (upper side of interface k). The angle δ_k is the angle by which the plane of incidence at a lower interface (interface k) has to be rotated to overlap with the plane of incidence at the next higher interface.

It is obvious that $\delta_1 = 0$ since the image ray emerges vertically at the surface and the x_s-axis was chosen to fall into the plane of incidence of 0_1. There is usually no need to have $\mathbf{R}_0$ expressed in the closed form of equation (4.63). For many applications, it will be sufficient to have $\mathbf{R}_0$ computed as the last step of a recursion. However, Hubral (1976a) describes one traveltime inversion method that is based on the closed-form expression (4.63).

4.3.8 Summary

A ray traced through a 3-D isovelocity-layer model with interfaces having arbitrary curvature permits the construction of a uniquely defined moving coordinate frame which moves with the expanding wavefront along the selected ray. Parameters pertaining to the ray permit the computation of a wavefront curvature matrix or its inverse, the radius matrix, at each ray point. The radius matrix provides a second-order approximation for the local shape of the wavefront and, if transformed into a principal coordinate system, determines the two principal radii of wavefront curvature. For any ray point, the matrix can be obtained recursively with the help of three laws expressed in the moving frame.

The second-order approximation to the wavefront determines the second-order terms of the traveltime observed at the ray emergence point. These analytic terms contain information about the subsurface that can be exploited for various purposes in forward and, in particular, inverse ray modeling.

4.4 Rays through specified end points

The recursive algorithms described above show how to trace a ray that starts in some arbitrary direction and how to compute the wavefront curvatures at any ray point.

A problem of special interest involves tracing the particular ray that passes through specified end points. There exists no universally accepted solution to this problem. However, a number of search schemes are available for determining such raypaths (Deschamps, 1972; Chander, 1977; Julian and Gubbins, 1977). We shall describe one search method which can be easily implemented with the help of the algorithms just described. The scheme is particularly useful if entire ray families for a common source (or receiver) point are to be traced and not just one ray alone.

A ray through specified points can be constructed by iteratively correcting the starting direction. Such iterative methods are known as shooting methods. We shall show that knowledge of wavefront curvature provides information that can be used to guide effectively the search toward the correct ray. Let us confine our attention to the case where a surface source S and a surface receiver G are to be connected by a ray (e.g., ray A' in Figure 4-5a). Other situations where ray end points are located arbitrarily within a velocity model can be dealt with similarly.

Let point $E^{(1)}$ be the ray emergence point (Figure 4-12) in a first attempt to trace a ray from S to G. Equation (4.47) provides a parabolic approximation for the true traveltime within the $[x_s, y_s]$ system at $E^{(1)}$:

$$t(x_s, y_s) = t(0,0) + \frac{\sin \beta_0}{v_1} x_s + \frac{1}{2v_1} \mathbf{X}_s \mathbf{S}_0 \mathbf{A}_0 \mathbf{S}_0 \mathbf{X}_s^T. \tag{4.64}$$

In equation (4.64), $t(0,0)$ is the traveltime from S to $E^{(1)}$. The $[x_s, y_s]$ system is determined by the emerging ray. The curvature matrix $\mathbf{A}_0$ is computed as a function of parameters along the ray between S and $E^{(1)}$. If (x_G, y_G) are the coordinates of G with respect to the $[x_s, y_s]$ system at $E^{(1)}$, then $t(x_G, y_G)$ provides a first approximation to the true traveltime between S and G. Equation (4.64) also gives an approximation for the direction of the desired ray emerging at G. Let us now demonstrate this point.

If the true traveltime were known in the vicinity of G, then equation (4.29) would provide the exact ray direction at G. If we, therefore, assume that the function $t(x_s, y_s)$ of equation (4.64) approximates this true traveltime sufficiently

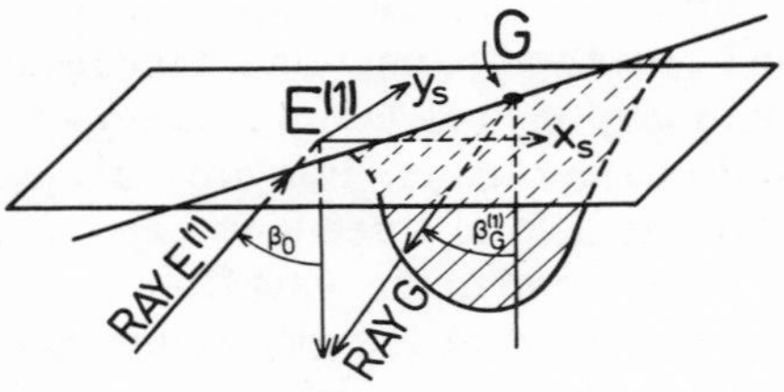

FIG. 4-12. Approximate ray at G obtained from the ray and curvature matrix of the emerging wavefront at $E^{(1)}$.

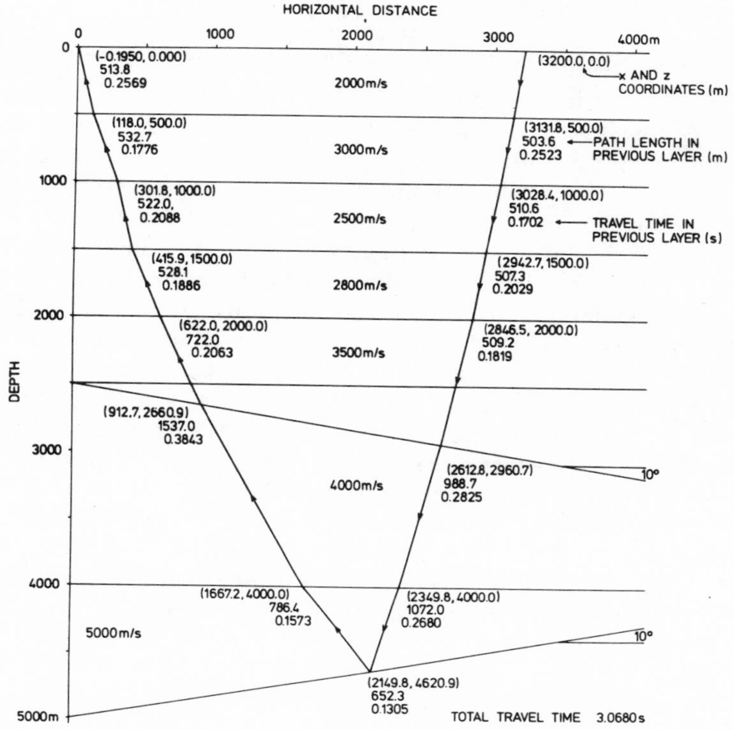

FIG. 4-13. Detailed parameter information along a computed raypath in a 2-D plane-dipping layer model (after Smith, 1978).

well around G, then we can use equation (4.64) in the gradient expression (4.29) to obtain an approximate direction of the ray emerging at G.

First of all, let us determine the direction of steepest ascent of $t(x_s,y_s)$ at G with respect to the $[x_s,y_s]$ system. It is designated by the unit vector $\mathbf{n}_G = \nabla t(x_s,y_s)/|\nabla t(x_s,y_s)||_G$ where

$$\nabla t(x_s,y_s)|_G = \left(\frac{\partial t}{\partial x_s}, \frac{\partial t}{\partial y_s}\right)\Bigg|_G = \left(\frac{\sin \beta_0}{v_1}, 0\right) + \frac{1}{v_1}\mathbf{X}_G \mathbf{S}_0 \mathbf{A}_0 \mathbf{S}_0. \tag{4.65}$$

The subscript 0 refers to the ray emergence point $E^{(1)}$, and $\mathbf{X}_G = (x_G, y_G)$.

The vertical plane through G that includes $\mathbf{n}_G$ is defined as the ray emergence plane (hachured in Figure 4-12) for the approximate ray at G. Within the $[x_s,y_s]$ system, it is determined by the following parametric equation

$$\mathbf{X}_s(q) = \mathbf{X}_G + q\,\mathbf{n}_G, \tag{4.66}$$

where q is the distance measured in direction $\mathbf{n}_G$ from G and $\mathbf{X}_s(q) = [x_s(q), y_s(q)]$.

Within the ray emergence plane, the approximate emergence angle $\beta_G^{(1)}$ is finally obtained from

$$\frac{\sin \beta_G^{(1)}}{v_1} = \left. \frac{\partial \{t[x_s(q), y_s(q)]\}}{\partial q} \right|_{q=0} . \qquad (4.67)$$

We now trace a ray in this direction within the vertical ray emergence plane back from G toward S. The resulting ray emerges at $E^{(2)}$, near the source S. By treating point S with respect to $E^{(2)}$ in the same way as point G with respect to $E^{(1)}$, we can establish a new $[x_s, y_s]$ coordinate system at $E^{(2)}$, find a second-order approximation to the true traveltime from G to S, and obtain an improved direction at S to start tracing a third ray from S back toward G.

Provided a reasonable first approximation to the desired ray can be found, successive approximations to the true traveltime and desired ray will converge much more quickly than in the more common practice where the last term in equation (4.65), i.e., the wavefront curvature term, is not considered. It should be mentioned that, as in other search schemes, convergence may be slow when the initial guess is poor (Ortega and Rheinbold, 1970). In fact, convergence is not even guaranteed for particular cases where the local second-order approximation for wavefront shape is inadequate.

A two-point ray-tracing scheme based on our approach was applied to the 2-D plane-dipping isovelocity layer model of Figure 4-13 (Smith, 1978) for a ray traced from $x = 3200$ m to $x = 0$. All essential parameters along the ray are marked for those readers who wish to test their own ray-tracing program.

4.5 Geometrical spreading

The subject treated in this section has little direct bearing on the problem of computing interval velocities from seismic reflection time curves. We have included it to demonstrate the close relationship between wavefront curvatures and geometrical spreading, the purely geometric change of amplitude associated with energy density along wavefronts. Readers concerned solely with computing interval velocities may wish to skip section 4.5. This section covers the purely geometrical aspects of the spreading function, and in addition it briefly describes its role in the context of obtaining dynamic solutions to the wave equation within the framework of the *asymptotic ray method.*

Our treatment is confined to 3-D isovelocity layer models. As a seismic pulse propagates through such an earth model, its amplitude generally decreases with increasing traveltime. A major part of the seismic attenuation is attributable to *geometrical spreading* (also called wavefront divergence); it can be estimated with the help of ray-tracing techniques. The subject is treated in detail in discussions of the asymptotic ray method [Cerveny et al, (1977)].

Presently, seismic reflection amplitudes often can be recovered with a high degree of precision. O'Doherty and Anstey (1971) and Sheriff (1975) give good accounts of the factors influencing amplitudes in horizontally layered media. Geometrical spreading is one of the dominant factors. O'Brien and Lucas (1971) were concerned with geometrical spreading in the study of well geophone arrivals. Newman (1973) discussed geometrical spreading for seismic reflections from a plane, horizontally layered subsurface. He derived an interesting correction formula for a zero-offset trace that makes use of the RMS velocity in scaling the amplitudes of primary reflections from a point source to their equivalents for vertically propagating plane waves. The formula can be exploited for the purpose of computing pseudosonic logs (Lavergne and Willm, 1977). It will be reviewed in section 4.5.2.

4.5.1 Divergence laws

Suppose that a spherically symmetric source of diverging acoustic waves is placed somewhere on the surface of the earth model or within an arbitrary layer (Figure 4-14). Then the effect of geometrical spreading can be determined solely from considering ratios of wavefront area, subtended by a given ray tube (Gutenberg, 1936). Simple "spherical spreading" is obviously inadequate for the assumed model because wavefronts are no longer spheres; rather, they are distorted by refraction, reflection, and interface curvature. As the amount of geometrical spreading differs for different portions of a wavefront, one has to select a specific ray (i.e., the central ray of a ray tube) in order to compute the divergence in its vicinity.

Consider a ray tube (Figure 4-14), including those rays which emerge within a narrow cone from source *S*. We shall now describe an algorithm for computing geometrical spreading by tracing the history of a small portion of the wavefront surface along the tube. For high-frequency signals, one can accept that energy which flows initially through the area ΔF_s on a unit sphere around the source will subsequently flow through the area ΔF_p at some arbitrary point **P** further

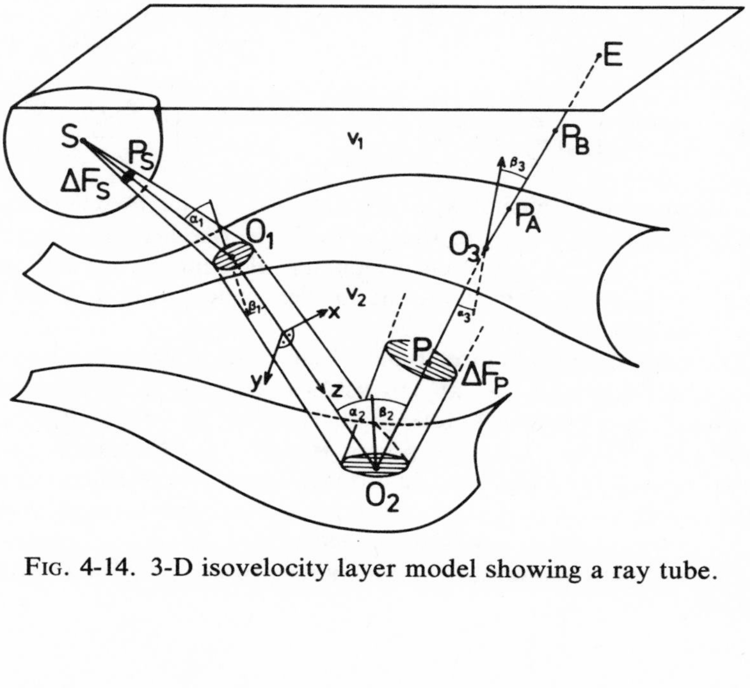

FIG. 4-14. 3-D isovelocity layer model showing a ray tube.

along the same ray. **P** may, of course, coincide with the position of the emerging ray at the surface. The intensity flow I_p through the ray tube at **P** relates to the intensity flow I_s through the ray tube at $\mathbf{P}_s$ by the following formula:

$$I_p\, dF_p = \lambda I_s\, dF_s, \tag{4.68}$$

where the proportionality factor λ accounts for all reflection and transmission losses that occur at intervening interfaces. These losses are well-described in books on the asymptotic ray method. dF_p and dF_s are the differential areas of the ray tube that result in the limit when the ray tube shrinks to zero. They are perpendicular to the central ray.

The differential quotient dF_p/dF_s assumes a finite value, which may be positive or negative. We shall now give an analytical expression for this quotient by considering a ray traced through a 3-D isovelocity layer model. The evaluation of dF_p/dF_s for subsurface velocity models more complex than considered here remains an interesting area of research. We mention here three recent publications on this subject (dealing with inhomogeneous velocity media and curved first-order interfaces) by Popov and Psencik (1976, 1978) and Hubral (1979a).

Amplitudes are proportional to the square root of intensities; therefore, the divergence factor d_p for point **P** is defined as

$$d_p = \left(\frac{dF_p}{dF_s}\right)^{1/2}. \tag{4.69}$$

Thus, d_p^{-1} gives the desired ratio of amplitudes. It can be either a real or imaginary quantity (see Cerveny et al, 1977). Note that we have ignored all reflection and transmission losses by setting the proportionality factor $\lambda = 1$ in equation (4.68).

In equation (4.69) it is usually assumed that dF_s pertains to the intersection of the unit sphere around S with the source ray tube traced to **P**. The problem of interest to us, namely the computation of the area ratio dF_p/dF_s, is much simplified by making use of the following equality, valid for the particular situation depicted in Figure 4-14.

$$\frac{dF_p}{dF_s} = \left(\frac{dF_{I,1}}{dF_s}\right)\cdot\left(\frac{dF_{T,1}}{dF_{I,1}}\right)\cdot\left(\frac{dF_{I,2}}{dF_{T,1}}\right)\cdot\left(\frac{dF_{R,2}}{dF_{I,2}}\right)\cdot\left(\frac{dF_p}{dF_{R,2}}\right). \tag{4.70}$$

All differential surface elements in equation (4.70) can be considered to be locally flat and perpendicular to the ray. $dF_{I,1}$ is the element of the incident wave at point 0_1; $dF_{T,1}$ is the corresponding element on the refracted side of point 0_1. $dF_{I,2}$ is the element on the incident side and $dF_{R,2}$ is the element on the reflected side of point 0_2, etc.

Expression (4.70) involves three types of geometrical changes of the ray tube along the selected ray. $dF_{T,k}/dF_{I,k}$ describes the change at point 0_k due to refraction and $dF_{R,k}/dF_{I,k}$ due to reflection. The quantity $dF_{I,k}/dF_{T,k-1}$ describes the change between two interface points separated by a straight ray segment through a homogeneous medium. In the following, expressions for the three types of ratios in equation (4.70) are referred to, respectively, as the refraction, reflection and transmission law of divergence.

It is obvious how equation (4.70) generalizes to account for the differential

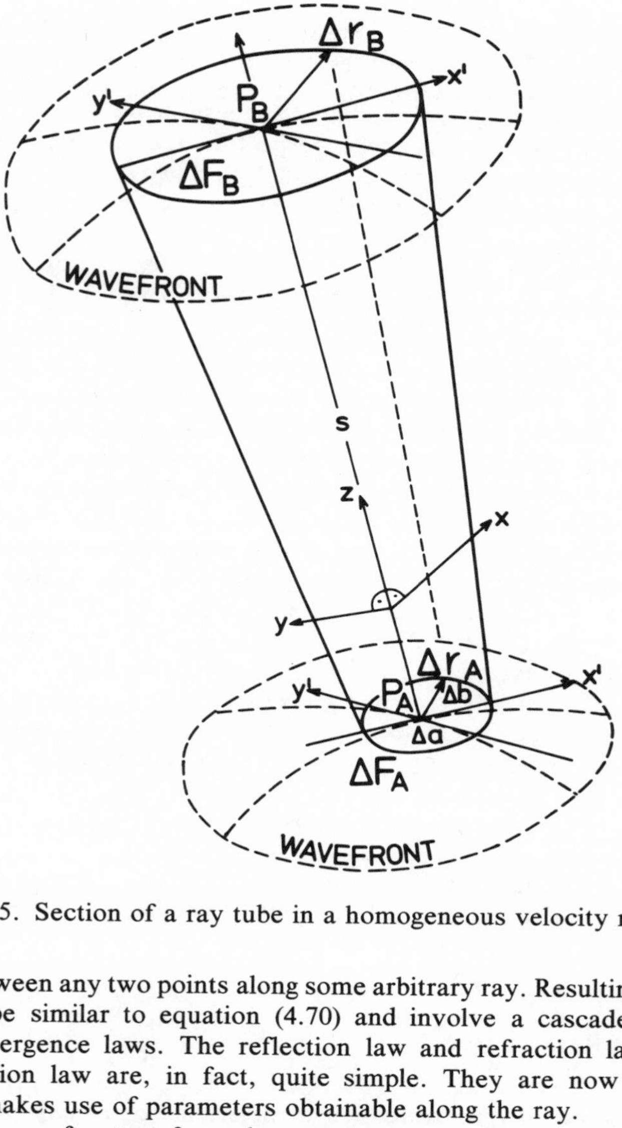

FIG. 4-15. Section of a ray tube in a homogeneous velocity medium.

area ratio between any two points along some arbitrary ray. Resulting expressions will always be similar to equation (4.70) and involve a cascaded product of the three divergence laws. The reflection law and refraction law as well as the transmission law are, in fact, quite simple. They are now expressed in a form that makes use of parameters obtainable along the ray.

As with a wavefront surface element, let us consider an interface surface element to be flat in the vicinity of the incident ray. Then the following equations, true for plane waves and planar interfaces, are also true for the differential area elements of curved wavefronts and interfaces;

$$dF_R = \frac{\cos \epsilon_R}{\cos \epsilon_I} \, dF_I, \tag{4.71}$$

and

$$dF_T = \frac{\cos \epsilon_T}{\cos \epsilon_I} \, dF_I. \tag{4.72}$$

ϵ_I is the incidence angle, ϵ_T the refraction angle, and ϵ_R the reflection angle.

The transmission law of divergence relates directly to the transmission law of curvatures. Let $\mathbf{P}_A$ and $\mathbf{P}_B$ be two ray points on opposite ends of a straight segment within a homogeneous medium. Suppose also that $\mathbf{R}_A$ and $\mathbf{R}_B$ are the radius matrices at these points with respect to the moving $[x, y, z]$ system. Figure 4-15 shows the wavefronts at $\mathbf{P}_A$ and $\mathbf{P}_B$ and two small elliptical wavefront area elements related to the same ray tube. In the limit where the thickness of the ray tube shrinks to zero, the area elements ΔF_A and ΔF_B shrink to zero as well. Now, the area ratio dF_B/dF_A is independent of the original area shapes ΔF_B and ΔF_A and the way the ray tube contracts. Therefore, we can choose any shape elements we wish for the computation of the desired differential area ratio. It is convenient to let them be small elliptical elements with axes pointing in the directions of the principal curvatures of the wavefront. As is shown in Appendix C, the desired ratio is given by

$$\frac{dF_B}{dF_A} = \frac{\det \mathbf{R}_B}{\det \mathbf{R}_A}. \tag{4.73}$$

The radius matrices $\mathbf{R}_A$ and $\mathbf{R}_B$ in equation (4.73) may, in fact, refer to any right-hand coordinate system at $\mathbf{P}_A$ or $\mathbf{P}_B$ which has the z-axis pointing in the direction of the ray. In one system, the principal coordinate system, the principal wavefront radii occur in the diagonal of the radius matrix. As principal radii can be positive or negative, the determinant of the radius matrix changes sign along a ray whenever one principal wavefront radius changes sign. Opposing signs may arise whenever a beam goes through a focal point or caustic. A seismic pulse of high-frequency content traveling along the ray then changes into its Hilbert transform. Naturally there exist focal points where both principal radii change sign. The seismic pulse passing through such a focal point then remains unchanged.

We have now established the three divergence laws that determine the geometrical spreading along a selected ray. The divergence factor (4.69) for point $\mathbf{P}$ of Figure 4-14 is consequently given by

$$d_p^2 = (\det \mathbf{R}_{I,1}) \cdot \left(\frac{\cos \beta_1}{\cos \alpha_1}\right) \cdot \left(\frac{\det \mathbf{R}_{I,2}}{\det \mathbf{R}_{T,1}}\right) \cdot \left(\frac{\cos \beta_2}{\cos \alpha_2}\right) \cdot \left(\frac{\det \mathbf{R}_P}{\det \mathbf{R}_{R,2}}\right). \tag{4.74}$$

The generalization of this expression to determine d_p at any point along any selected ray is evident. The radius matrices required to evaluate equation (4.74) are obtained by the recursive method described in section 4.3.

Equation (4.74) involves only parameters encountered along the ray. The divergence factor d_p thus accounts for all geometrical focusing and defocusing effects which can result from the presence of curved interfaces. It takes on much simpler analytic forms in special cases that will be discussed later. We want to stress, however, that frequently divergence factors need not be expressed in closed form. Typically, it is more convenient to apply the divergence laws recursively along with the ray tracing and application of the curvature laws, while following the progressively expanding (or contracting) wavefront.

The above divergence laws can be extended to inhomogeneous media with curved first-order interfaces. While the refraction and reflection laws remain

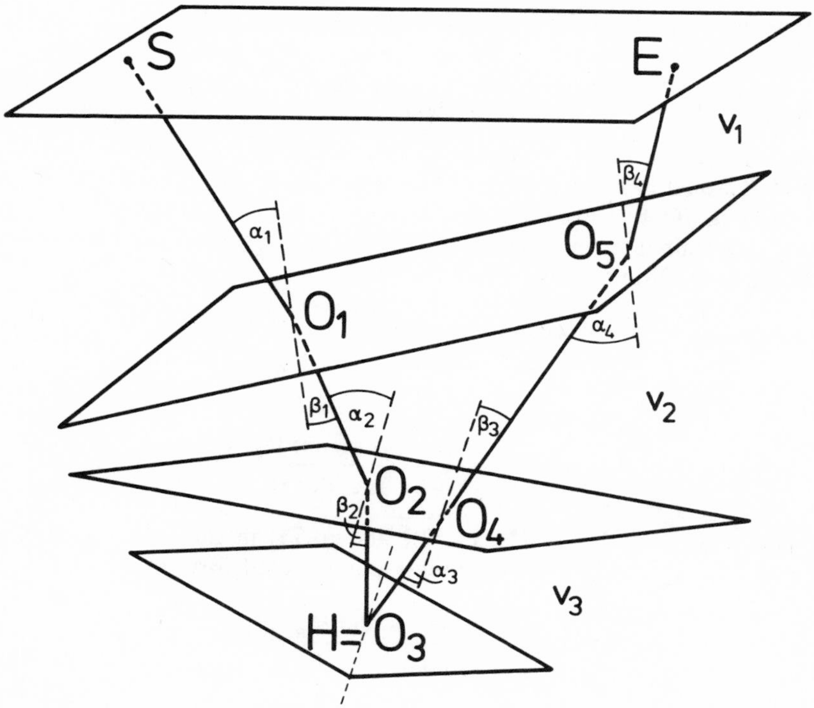

FIG. 4-16. 3-D plane isovelocity-layer model showing a primary reflected ray.

the same, the transmission law takes on a different form that can also be based on the evaluation of wavefront curvatures (Kline, 1961; Kline and Kay, 1965, p. 184–186; Hubral, 1979a).

4.5.2 Recovering divergence factors from traveltimes (plane-dipping layer case)

Just as we are able to recover the wavefront curvature matrix of an emerging wavefront from traveltime measurements, we might ask if we can also recover the divergence factor. As a matter of fact, it is not always necessary to compute the divergence factor as a function of parameters along a ray. In some situations it can be determined from traveltime measurements.

Figure 4-16 shows a 3-D plane, isovelocity-layer model with interfaces of different dip and strike. A surface source S and a surface point E are connected by a ray of a primary P-wave reflected at point H on the Nth interface (Figure 4-16 shows H on the third interface). In this case, the refraction law (4.38) reduces to

$$\mathbf{R}_T = \frac{v_I}{v_T}\mathbf{D}^{-1}\mathbf{S}^{-1}\mathbf{R}_I\mathbf{S}^{-1}\mathbf{D}, \tag{4.75a}$$

and the reflection law (4.39) becomes

$$\mathbf{R}_R = \frac{v_I}{v_R} \mathbf{D}^{-1}\mathbf{I}_R \mathbf{S}'^{-1}\mathbf{R}_I \mathbf{S}'^{-1}\mathbf{I}_R \mathbf{D}. \tag{4.75b}$$

$\mathbf{R}_I$ is the radius matrix on the incident side of an interface, $\mathbf{R}_T$ on the refracted side, and $\mathbf{R}_R$ on the reflected side. The matrices $\mathbf{S}$, $\mathbf{S}'$, $\mathbf{I}_R$, and $\mathbf{D}$ are defined as in equations (4.38) and (4.39).

Taking determinants on both sides of equations (4.75a) and (4.75b) leads to the following conditions

$$\frac{\cos \epsilon_I}{\cos \epsilon_T} = \frac{v_I^2 \cos \epsilon_T \det \mathbf{R}_I}{v_T^2 \cos \epsilon_I \det \mathbf{R}_T}, \tag{4.76a}$$

and

$$\frac{\cos \epsilon_I}{\cos \epsilon_R} = \frac{v_I^2 \cos \epsilon_R \det \mathbf{R}_I}{v_R^2 \cos \epsilon_I \det \mathbf{R}_R}. \tag{4.76b}$$

The divergence formulas (4.71), (4.72), and (4.73) in the 3-D plane-layer case can hence be expressed in the seemingly complicated form

$$dF_R = \frac{v_R^2 \cos \epsilon_I \det \mathbf{R}_R}{v_I^2 \cos \epsilon_R \det \mathbf{R}_I} dF_I, \tag{4.77}$$

$$dF_T = \frac{v_T^2 \cos \epsilon_I \det \mathbf{R}_T}{v_I^2 \cos \epsilon_T \det \mathbf{R}_I} dF_I, \tag{4.78}$$

and

$$dF_B = \frac{\det \mathbf{R}_B}{\det \mathbf{R}_A} dF_A. \tag{4.79}$$

However, when these three laws are applied recursively to a ray such as that shown in Figure 4-16, the expression for the divergence factor takes on a simple closed analytic form

$$d_E = \left[\det \mathbf{R}_E \prod_{k=1}^{2N-2} \left(\frac{\cos \alpha_k}{\cos \beta_k}\right)\right]^{1/2}. \tag{4.80}$$

[Note that for the computation of equation (4.80) we have not allowed mode-converted waves.]

The square of the right side of equation (4.80) is directly proportional to the determinant of the radius matrix $\mathbf{R}_E$ of the emerging wavefront. $\mathbf{R}_E$ can be recovered from reflection time measurements in three profiles through E [see formula (4.47)]. The factor by which the determinant is multiplied includes only cosines of incidence and refraction angles along the ray and reduces to unity when E coincides with S.

Therefore, one can compute the divergence factor of the reflected wave at S entirely from traveltime measurements. The only assumption we have to make about the velocity model is that all interfaces are planar. Furthermore, if all layers are horizontal, then $\det \mathbf{R}_{E=S} = R_0^2$, where R_0 is the (azimuth-independent) radius of wavefront curvature of the vertically emerging wave at S. For this special case, one consequently obtains $d_{E=S} = R_0$. This result is also given by Newman (1973) who shows that the divergence factor relates to the RMS velocity and is thus available from conventional CDP stacking-velocity analysis:

$$d_{E=S} = R_0 = \frac{t(0)}{v_1} V_{\mathrm{RMS}}^2 . \tag{4.81}$$

This relation between R_0 and V_{RMS} was given in Example 1 in section 4.3.7. For further details on the application of formula (4.81) for the inversion of seismic traces, see Müller (1971).

For a separated surface source and surface receiver above a horizontally layered medium, symmetry of incidence and refraction angles in equation (4.80) leads to

$$d_E^2 = \det \mathbf{R}_E . \tag{4.82}$$

$\mathbf{R}_E$ can again be obtained from traveltime measurements on the surface of the earth. The divergence factor can thus be computed from traveltimes without actually performing ray tracing. In fact, the velocity distribution need not be known. Note that due to the circular symmetry of traveltimes, one profile connecting S with E is sufficient for computation of d_E.

With the help of wavefront curvature laws, one obtains for $\mathbf{R}_E$

$$\mathbf{R}_E = \begin{bmatrix} R_x' & 0 \\ 0 & R_y' \end{bmatrix}, \tag{4.83}$$

with

$$R_x' = \frac{2}{v_1} \sum_{j=1}^{N} v_j s_j \frac{\cos^2 \alpha_1}{\cos^2 \alpha_j}; \; (\alpha_N = \beta_{N-1}), \tag{4.84}$$

and

$$R_y' = \frac{2}{v_1} \sum_{j=1}^{N} v_j s_j . \tag{4.85}$$

R_x' is the radius of the emerging wavefront at E within the vertical plane containing the emerging ray. R_y' is the radius within the orthogonal plane which also includes the emerging ray. Both R_x' and R_y' are principal radii of curvature. s_j is the ray segment length in the jth layer. Consequently, for horizontally layered media expression (4.82) can be expressed simply as $d_E^2 = R_x' R_y'$.

4.5.3 Asymptotic ray method (a brief review)

The asymptotic ray method, also called simply the ray method, is occasionally considered in exploration seismology as a tool for forward dynamic wave-equation modeling (Shah, 1974; Smith, 1977; May and Hron, 1978; Schneider, 1979). With respect to its application in refraction seismology and deep seismic sounding, the method is well described by Cerveny and Ravindra (1971) and Cerveny et al (1977).

In the ray method, the overall wave field is decomposed into individual "elementary waves"—one for each ray. Many rays may have to be traced between a specified source and a receiver in order to construct the total wave field by superposition of elementary waves. Elementary waves (i.e., the pulses along considered rays) can be expressed in terms of source parameters and model parameters found along the traced rays.

The ray method provides an interesting and economical alternative to various integral and finite difference schemes. Its most significant qualities are its simplicity and its intuitive appeal. The method is reviewed only briefly here for models assumed to have acoustic properties.

As we have been considering only kinematic problems up to this point, we have shown no concern for the density distribution in our (usually) thick-layer models. Without disturbing our traveltime considerations, density layering could have been quite complex and caused more reflections than those assumed to result from first-order velocity boundaries. For the computation of ray amplitudes for elementary waves by the ray method, we must make the additional severe assumption that the thick constant velocity layers are also layers of constant or smoothly changing density. With this assumption, we may neglect many possible first-order discontinuities in the density field that actually exist in the earth, with consequent degradation of the generality and accuracy of derived dynamic wave properties. Recall our statement made previously: amplitudes are much more affected by thin layering than are traveltimes.

In the literature of the late 1950s, it was shown that practically all solutions for seismic body (and head) waves in thick (isotropic) layers by various wave equation methods can equally well be approximated by the so-called ray method. This method provides dynamic solutions for a very general class of subsurface models, many of them not yet tractable by other methods. Such solutions, however, are obtained at the expense of certain limitations. For instance, the ray method fails to provide exact solutions at caustics, cusps, and at points where waves are critically reflected and head waves are generated.

The basic wave equation describing pressure $p(x_0, y_0, z_0, t)$ in an acoustic inhomogeneous medium of velocity $v(x_0, y_0, z_0)$ is

$$\left(\frac{\partial^2}{\partial x_0^2} + \frac{\partial^2}{\partial y_0^2} + \frac{\partial^2}{\partial z_0^2}\right) p - \frac{1}{v^2(x_0, y_0, z_0)} \frac{\partial^2 p}{\partial t^2} = 0.$$

For a harmonic source of frequency ω, the asymptotic ray series solution of an elementary wave in such a medium can be written as:

$$p(x_0, y_0, z_0, t) = \sum_{n=0}^{\infty} A_n(x_0, y_0, z_0)\, e^{i\omega\,[t - \tau(x_0, y_0, z_0)]}\, (i\omega)^{-n},$$

where $i = \sqrt{-1}$. The coefficients $A_n(x_0, y_0, z_0)$ are the amplitude terms of the ray series. The equation describing the position of a selected wavefront at any time t is $t = \tau(x_0, y_0, z_0)$. Expressions for the phase function τ and the terms A_n $(n = 0, 1, \ldots)$ are obtained by substituting the ray series into the wave equation. The result is that τ is a solution of the eikonal equation $(\nabla \tau)^2 = 1/v^2$, and the coefficients A_n, $(n = 0, 1, \ldots)$ are solutions of a recursive system of differential equations, the so-called transport equations. Both A_n and τ may be complex quantities.

For a high-frequency source, the ray series terms for $n \geq 1$ rapidly approach zero, and one obtains the geometric optics solution by evaluating only the leading term in the expansion,

$$p(x_0, y_0, z_0, t) = A_0(x_0, y_0, z_0) e^{i\omega [t - \tau(x_0, y_0, z_0)]}.$$

This solution, also called the zero-order solution, is often sufficient for seismic body waves. It has an intuitive appeal and can be obtained easily following two simple principles.

(1) Energy propagates along rays and does not flow through the side-walls of ray tubes.

(2) At ray-interface intersections, locally occurring amplitude changes (for the reflected or refracted waves) can be treated like those that result when curved wavefronts and interfaces are replaced by tangent plane waves and tangent plane interfaces (principle of isolated elements).

Note that the zero-order solution is valid only for wavelengths that are short in comparison with the thicknesses of layers and radii of curvatures of the first-order interfaces.

Curved wavefronts of high-frequency signals thus can be treated locally like plane waves. To satisfy the first principle, one computes the geometrical spreading factor. To satisfy the second principle, one computes reflection and refraction coefficients using the familiar equations of the Knott-Zoeppritz type (Zoeppritz, 1919; Cerveny and Ravindra, 1971). More on the asymptotic ray method in acoustic media is found in Ahluwalia and Keller (1977).

4.5.4 Summary and perspectives

Geometrical spreading, like the change in curvature of a wavefront that travels from a source to a receiver through a layered earth, can be expressed solely in terms of seismic parameters encountered along the ray. Geometrical spreading accounts for focusing and defocusing effects that can sometimes be observed on real seismic sections. Wavefront curvatures and ray directions of emerging waves are contained in the first- and second-order derivatives of traveltime curves.

For exploration situations, in which it can be assumed that the earth model consists of plane isovelocity layers, the divergence factor in the reflection data can be inferred from observed traveltime curves. One might hope to relate the divergence factor of a reflected primary wavefront, along a normal ray, to the NMO velocity of a CDP-profile (see section 5.1.2) which, as we will see, can be also expressed in terms of parameters along the normal ray.

Divergence factors for reflected primary waves of a coincident source-receiver pair include effects of the curvature of reflectors, whereas the NMO velocity of a primary CDP reflection is not influenced by the curvature of the reflector to which it pertains (see section 6.2). Therefore, layered models having curved reflectors must be excluded if we hope to find a relationship between the NMO velocity and the divergence factor of a primary reflection.

An essential feature of the ray method is that the forward dynamic modeling problem is solved *after* the forward kinematic ray modeling solution has been obtained. At little extra expense, one can therefore obtain the so-called zero-order solution for a specified source wavelet and assumed layer densities. This solution, exact in the high-frequency limit, is based largely on computing the divergence factor along a ray.

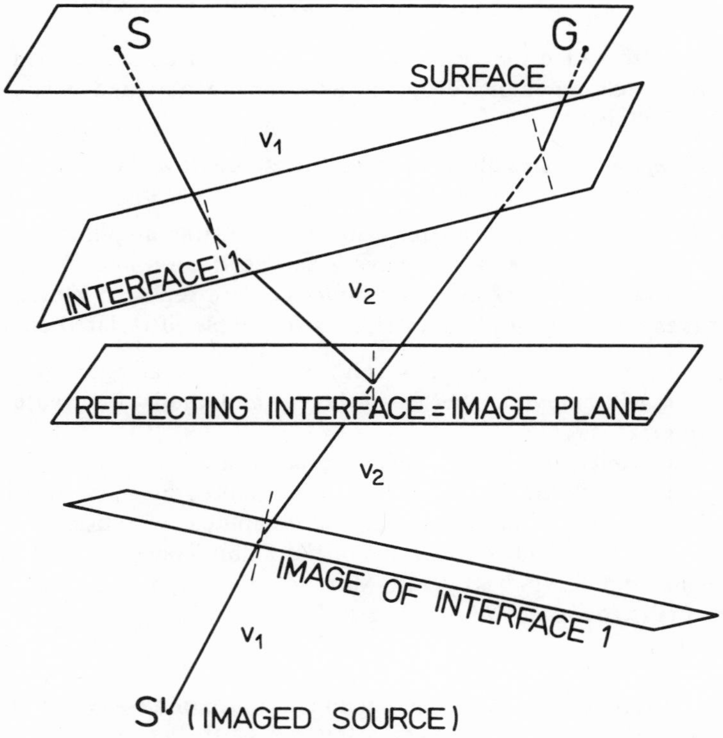

FIG. 4-17. Plane interface imaging. The source and downgoing ray are mirrored with respect to the image plane.

In conclusion, let us also shed further light on the principles of plane interface imaging. As a reflecting plane within an isovelocity layer model acts as a nondistorting planar mirror for nonconverted waves, one can ignore it by replacing the original model by an equivalent one possessing identical properties at all mirror points with respect to the reflecting plane.

Rather than following reflected raypaths in the original model, one can thus follow nonreflected paths in the equivalent model in an upward or downward direction only. Upward and downward in this context means that the ray never passes more than once through the same (planar) interface of the equivalent model. The various *reflection laws* governing ray direction, wavefront curvatures, or divergence then become unnecessary and many solutions can often be formulated in a simpler or more compact analytic form.

Figure 4-17 shows a 3-D plane isovelocity layer model with interfaces of different dips. A point source is placed at *S*. This source is mirrored at the second horizon, reflector for the considered *P*-wave. Traveltimes are the same in both the original model and in the equivalent model. Imaging the source involves tracing rays in an upward direction through the equivalent model. Conversely, imaging the receivers involves rays that need only be traced downward.

5 Common-datum-point (CDP) methods

The ray-theoretical concepts as presented in chapter 4 are little tailored to the demands of recording techniques now standard in the seismic reflection method. All ray-theoretical considerations from this chapter on will be put in the context of the common-datum-point (CDP) technique.

The introduction of the CDP method meant a major breakthrough for seismic exploration (Mayne, 1962, 1967). CDP data provide redundancy of information for use in attenuating multiples and random noise as well as for estimating the subsurface velocity distribution. Continuous CDP profiling, now standard practice all over the world, offers numerous advantages over other seismic profiling techniques previously used. In this chapter, we will first review commonly used definitions for the conventional 2-D CDP technique and then extend these to include modern 3-D CDP profiling methods.

5.1 2-D data gathering

5.1.1 Seismic profile

A *2-D seismic profile* or *spread* consists of a single shot location and a series of (generally equispaced) receiver locations along a straight line. Each receiver location is normally defined as the center of an array of geophones or hydrophones. The distance from a shot to a receiver is called the *(shot-receiver) offset;* more particularly, the distance to the nearest receiver location is called the *near-trace offset* or *in-line offset,* the distance to the farthest receiver location is called the *maximum offset,* whereas the distance from the first to the last receiver location is called the *spread length.*

When the spread of receivers for a given shot straddles that shot (i.e., some receivers are forward of the shotpoint and some are behind), we speak of *split-spread shooting.* In contrast, when all the receivers are on one side of a shot, we use the term *single-ended shooting* or *off-end shooting.* A land seismic survey can involve either or both types of shooting. Typical marine surveys involve a streamer cable towed behind a shooting boat, and thus are almost invariably shot off-end.

Seismic traces arranged in the order of receivers for a given shot comprise a *seismic record* (one record for each seismic profile). Figure 5-1 shows, for two different subsurface models, rays and traveltimes to a reflector in a split-spread record (Figure 5-1a) and those in an off-end record (Figure 5-1b). Whether the layers are bounded by plane, horizontal interfaces or by interfaces that are curved in 3-D space, both primary and multiple reflections (symmetric and asymmetric) in a seismic record have nearly hyperbolic *traveltime curves.*

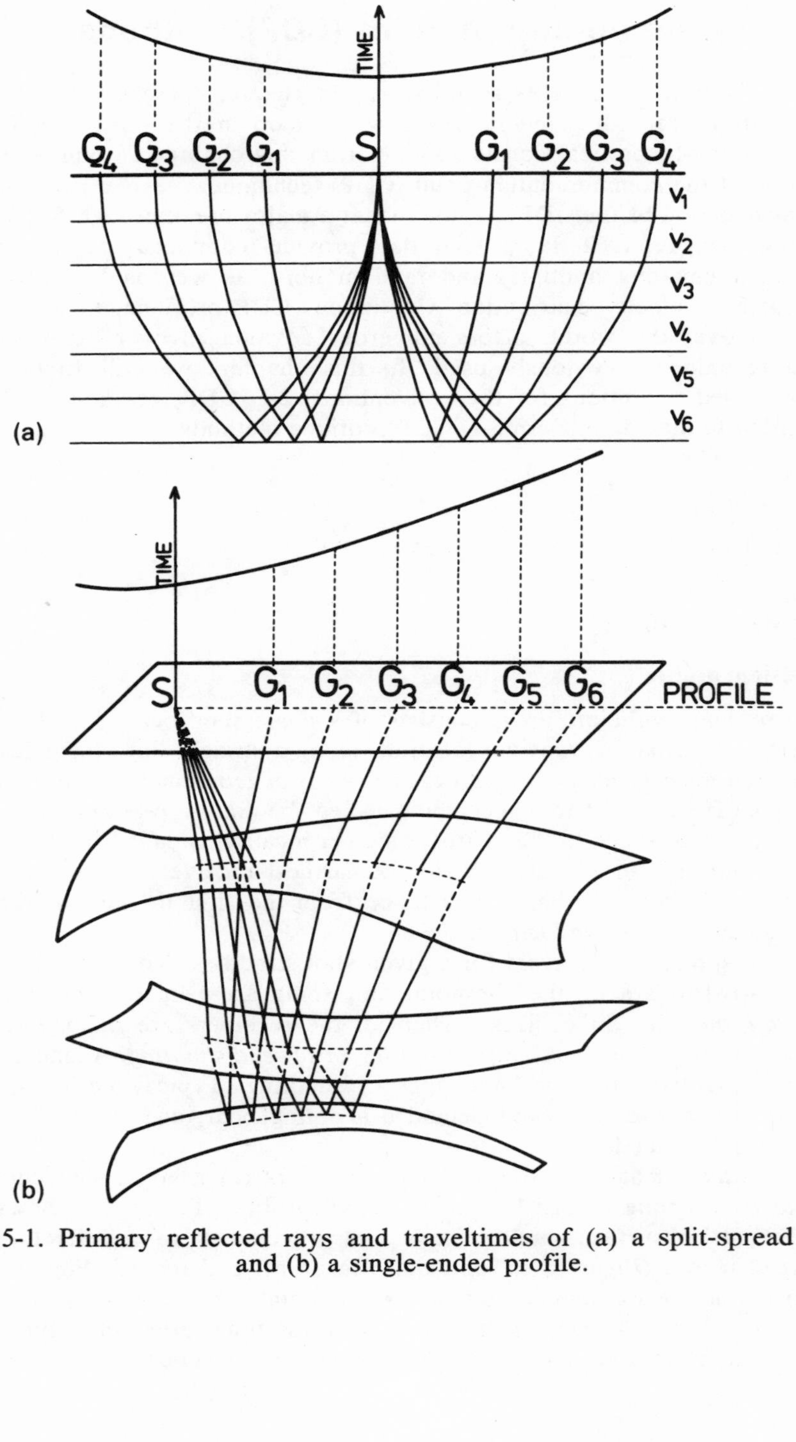

FIG. 5-1. Primary reflected rays and traveltimes of (a) a split-spread profile and (b) a single-ended profile.

Their apexes are generally not at *zero offset.* A sufficient condition for the apex to occur at zero offset is that the reflecting beds and the isovelocity layers in the subsurface be horizontal. The hyperbola-like traveltime curves for primaries and multiples reduce to exact hyperbolas only if the subsurface consists of a constant-velocity medium above a plane reflector of arbitrary dip and strike. We refer to such a model as a homogeneous plane-layer model (see Levin, 1971).

5.1.2 CDP profile

A *datum point* is defined as the midpoint between a shot and receiver. A *common datum point profile* consists of an equal number of shots and receivers placed symmetrically on a straight line about a common profile midpoint or common datum point, CDP (Figure 5-2). For each shot position, signals are recorded at the receiver position symmetrically opposite the shot position on the other side of the CDP point. The traces recorded in a CDP profile constitute a *CDP family* or *CDP record.* The number of shots (or receivers) in a CDP family determines the (subsurface) coverage. A CDP family with N shots provides an N-fold CDP coverage and is said to cover the ground by $N \times 100$ percent. Commonly used CDP coverages are 6-, 12-, 24-, 48-, and 96-fold, but other multiplicities, as 8-, 10-, 15-fold, etc., can be considered with modern instruments.

For simplicity in theoretical considerations, we often need the idea of zero-offset single subsurface coverage involving fictitious coinciding shot-receiver pairs uniformly densely covering a *seismic line.* If a CDP profile is placed along the surface of a 3-D subsurface model (Figure 5-2), then all reflection points (e.g., of a primary reflection) belonging to the various shot-receiver pairs generally span only a small portion of the considered reflector. This portion shrinks to zero for horizontally layered models. Rays then pass through a common depth point (Figure 5-3b), which falls vertically below the CDP.

Many exploration seismologists continue to refer to a CDP profile as a common-depth-point (also CDP), common-reflection-point (CRP), or common-ground-point (CGP) profile. In order to have the name for the profiling technique be model-independent, we honor the abbreviation CDP and imply that it stands for common datum point. This proposed name is synonymous with names like common-surface-point (CSP) and common midpoint (CMP), the latter two abbreviations being less frequently used.

The relationship between a CDP family and a CDP profile is equivalent to the relationship between a seismic record and a seismic profile. A CDP family offers basically two advantages (over a seismic record) in detecting and discriminating primary and multiple reflections—tasks of fundamental importance in computing velocity analyses for CDP stacking and for solving the general inverse seismic problem:

(1) CDP reflection times (for both primaries and symmetric multiples), plotted as a function of offset, fall upon one-sided hyperbola-like curves with apexes at zero offset. These traveltime curves are called *CDP reflection time curves.* Their apex positions coincide with zero offset because the first derivative of a CDP reflection time curve with respect to offset is zero at this point.

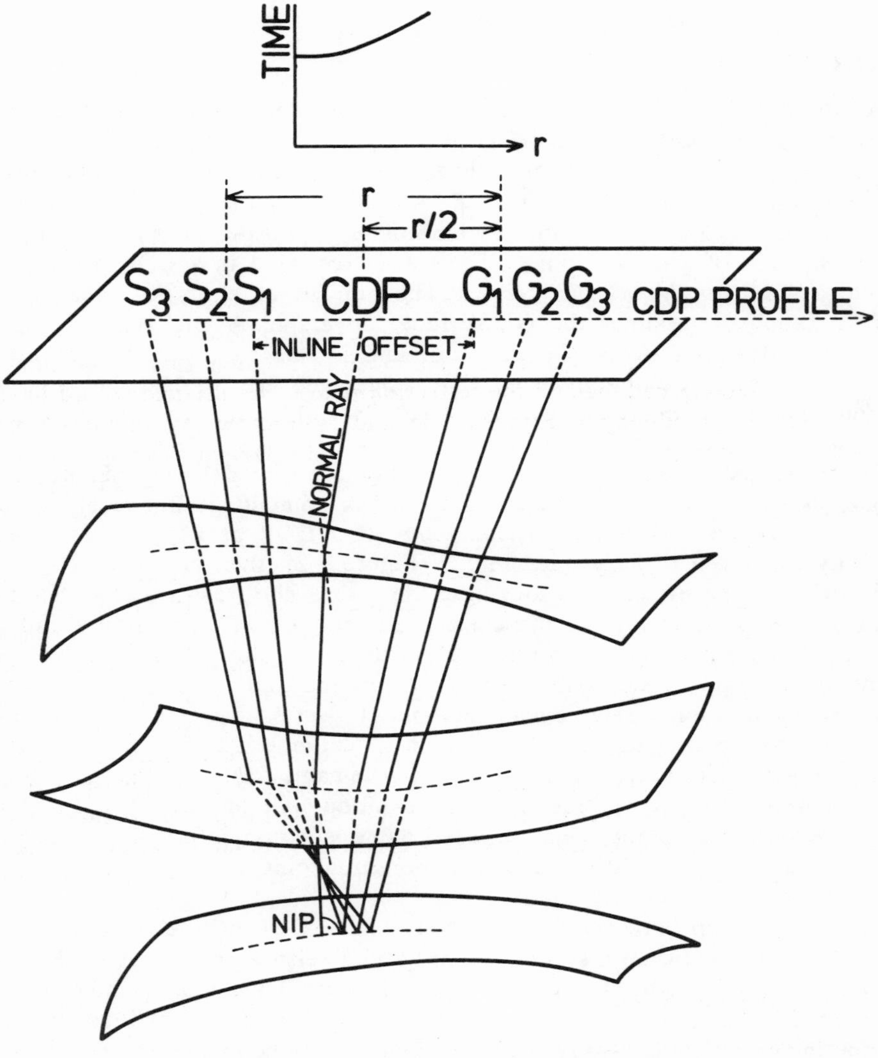

FIG. 5-2. Primary reflected rays of a CDP profile.

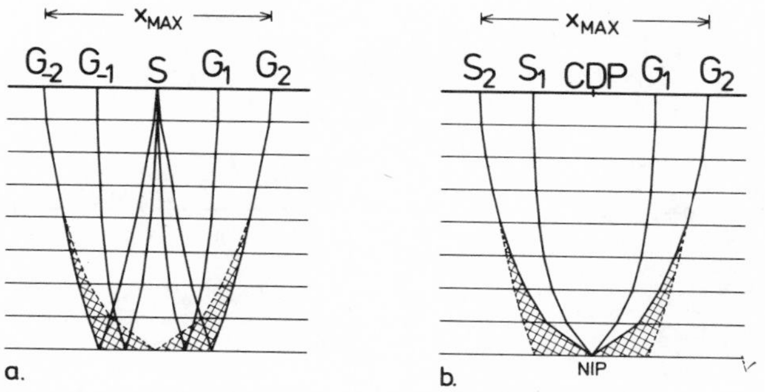

FIG. 5-3. Profiles of equal length (a) a split-spread profile and (b) a CDP profile.

(2) Reflections in a CDP family provide more "pinpointed" or "focused" information about the subsurface than do reflections in a seismic record because the CDP reflection points cover a much smaller portion on a reflecting bed than do the reflection points of a seismic record having equal spread-length. Figures 5-3a and 5-3b show ray families for an equally long symmetrical split-spread and CDP profile over identical subsurface models.

Both ray families cover a certain area between the surface and the reflector. This area is larger for split-spread rays than for CDP rays. The additional subsurface area traversed by split-spread rays on their way to a chosen reflector is hachured. For the CDP family, all rays reflect from the common depth point NIP. By replacing the layers of Figure 5-3 by a single constant-velocity layer, the difference in area is easily established; split-spread rays traverse one-and-a-half times the area covered by CDP rays.

From this simple observation, one can deduce that irregularities in the deeper subsurface enter to a higher degree into split-spread and off-end seismic records than into CDP records of equal spread-lengths. Near-surface anomalies within the earth, unfortunately, affect the timing of reflections in CDP records just as much as they do in seismic records.

5.1.3 Continuous profiling

Continuous profiling results from moving a seismic profile by equal steps along a *seismic line* in a more or less straight direction. The steps must be small enough to ensure at least a gapless one-fold coverage. If multiple coverage is applied, successive CDP families are constructed from successive seismic records by rearranging the traces along the line.

With the help of an example, let us elaborate on this important procedure of obtaining CDP families (or CDP gathers) from sequential seismic records. Figure 5-4 shows a single-ended spread (profile 1) with one shot position (cross) and $N = 6$ receiver positions (circle). The *receiver spacing* is Δx, and the near-trace

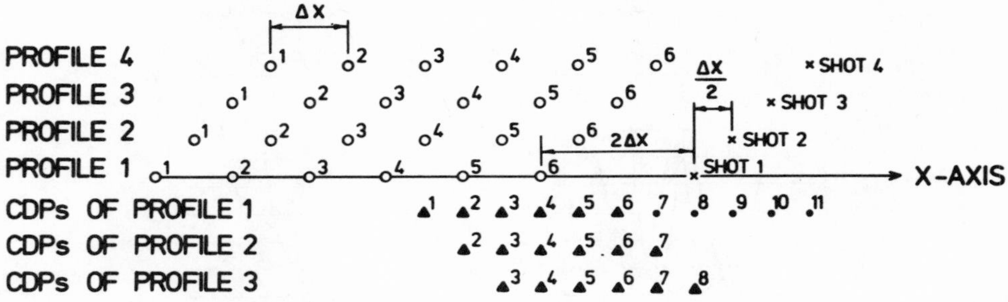

FIG. 5-4. Continuous CDP profiling.

offset is $2\Delta x$. This single-ended spread includes N one-fold covered CDP profiles with different offsets. The CDP positions (triangles) for profile 1 are placed in the first row below the x-axis (which coincides with the seismic line). CDP spacing is always half the receiver spacing.

If profile 1 is shifted in increments of $\Delta x/2$ to the right, as is often the case in marine work, the seismic profiles 2, 3, ... result. Though all seismic profiles are, in fact, confined to the x-axis, for tutorial reasons they are here placed in successive rows above profile 1 in addition to being shifted to their proper lateral positions. All CDPs of profile 2 are placed below the CDPs of profile 1, etc. Two seismic off-end profiles displaced by $\Delta x/2$ along the seismic line have $N - 1$ common datum points in common. These CDPs are two-fold covered. N seismic profiles successively displaced by the *shot spacing* $\Delta x/2$ are needed to build up N-fold CDP coverage from the Nth CDP location onward. If seismic spreads are displaced by the receiver spacing (or twice the CDP spacing) Δx, then the maximum coverage that can be obtained is half the number of receiver locations in the cable. If seismic spreads are displaced by $2\Delta x$, then the CDP coverage is a quarter of the number of receivers, etc. Summarized are four general statements about CDP-gathered data:

(1) In-line offset does not affect the multiplicity of CDP coverage. It influences only the positions of CDPs.

(2) For shot spacing greater than or equal to half the receiver spacing, the CDP spacing is always half the receiver spacing. For each common datum point, one obtains one CDP family.

(3) The shot spacing must equal half the receiver spacing to achieve maximum CDP coverage (equal to the number of receivers in the spread).

(4) Doubling the shot spacing, not exceeding $N\Delta x/2$, halves the CDP coverage and the number of seismic records without affecting the number of resulting CDP families.

From Figure 5-4 one can deduce the so-called *gather chart* describing the reordering of traces from successive seismic records into CDP gathers (i.e., CDP families) for successively numbered CDPs. The gather chart for the configuration of Figure 5-4 ($N = 6$, $\Delta x = 100$ m, and shot spacing $\Delta x/2$) is shown in Table 5-1. (n, m) in the chart denotes the nth shotpoint (or seismic record) and the mth receiver (or trace).

5.2 3-D Data gathering

In recent years, advancements in seismic data acquisition and processing capabilities have brought on renewed interest in the third dimension in seismic data gathering, processing, and interpretation. Conventional 2-D CDP profiling techniques along straight or nearly straight lines have the disadvantage that they do not permit discrimination of reflections from actual locations within the 3-D space of the subsurface. Thus, CDP profiling has been modified and extended in various ways to account for 3-D aspects of the subsurface.

5.2.1 Strip shooting

A particular, but not truly adequate, approach to the 3-D seismic method consists in adhering basically to line shooting while striving additionally to obtain information on crossdip, the component of dip in the direction perpendicular to the seismic line of shooting. The primary purpose here is to obtain *along a line* estimates of the *total* gradient ∇T_0 of the zero-offset reflection time $T_0(x_0, y_0)$ of primary reflections with respect to the horizontal coordinates x_0, y_0 on the surface of the earth. Later we shall see that ∇T_0 is needed for the determination of interval velocities for 3-D dipping layers or for construction of depth maps by time-to-depth migration. Direct measurement of ∇T_0 is often preferable to estimating this vector from T_0 time maps as in classical practice. (For economical reasons T_0 time maps are typically obtained from rather coarse grids of crossing seismic lines.)

Various field methods have been attempted for estimating crossdip. We will not describe them in detail here. They all provide a certain scattering of the datum points of shot-receiver pairs to one or both sides of a fictitious or

Table 5.1. Gather chart.

CDP							COVERAGE
1	(1,1)						1
2	(2,1)	(1,2)					2
3	(3,1)	(2,2)	(1,3)				3
4	(4,1)	(3,2)	(2,3)	(1,4)			4
5	(5,1)	(4,2)	(3,3)	(2,4)	(1,5)		5
6	(6,1)	(5,2)	(4,3)	(3,4)	(2,5)	(1,6)	6
7	(6,1)	(5,2)	(4,3)	(3,4)	(2,5)	(1,6)	6
8	(6,1)	(5,2)	(4,3)	(3,4)	(2,5)	(1,6)	6
.							
.							

hypothetical seismic line and are, therefore, categorized as "strip profiling." Common datum points in the sense described in the previous section are replaced, to a large extent, by an areal scatter of nearly common datum points comprised of those datum points having an approximately common coordinate along the fictitious seismic line.

In several versions of strip profiling, a straight line of receivers is maintained while the shot- or vibration points are located on both sides of the line, either in regular, irregular, or somewhat random patterns. In Vibroseis® surveys, it is operationally preferable to have the vibrator points located along a straight line while geophones are planted on both sides of the line.

In other approaches to strip profiling, geophones and shots are located on the same nonstraight line. The line might run in a zigzag manner generally along the fictitious seismic line preferably with small angles (<30 degrees) between actual zigzag lines and the seismic line. A feature of zigzag lines is the alternation of points from coincidence with genuine CDP (in the middle of straight elements of the zigzag line) to points having a maximum range of lateral displacements of datum points (at the corners of the zigzag line). The latter are points where crossdip can best be estimated, whereas the CDPs are locations where the parameters of CDP traveltime curves can best be determined. Often, land surveys are constrained by terrain or existing road networks to follow meandering and crooked lines rather than regular zigzag lines.

Strip profiling is usually limited to land surveys or surveys in shallow water where lines of hydrophones can be fixed in place. It is not well suited for conventional towed-cable surveys, though attempts have been made to circumvent the restrictions inherent in the marine seismic one-boat method. For instance, strong cross currents have been used to advantage, especially tidal currents. Furthermore, airguns at appropriate lateral offsets at both sides of the boat have been fired alternately.

A common feature of strip shooting methods is the restriction of CDP-traveltime information to predominantly one azimuth only; i.e., that of the fictitious seismic line. This restriction is not necessarily a disadvantage in practice because the vectors from shot locations to receiver locations do not deviate from the general direction of the fictitious seismic line.

5.2.2 Areal seismic surveys

3-D seismics in a more strict meaning involves a rather uniform coverage of datum points over the entire survey area of interest. Where possible, a regular grid of datum points is most desirable. One way to obtain such a grid is simply to shoot many closely spaced, parallel straight lines. To obtain a square grid of datum points in this manner, we would have the spacing between lines be half the in-line receiver group interval.

In order to produce a specified total number of datum points having a desired multifold coverage, a trade-off can be made between the number of shotpoint locations occupied and number of receiver channels recorded. For land surveys, economic considerations in field operations usually favor minimizing the number of shotpoint or vibrator-point locations while recording a large number of channels

® Registered trade and service mark of Conoco.

(96, 192, or more). Given this trade-off, it is often preferable to conduct 3-D land surveys in overlapping, parallel strip profiles. Uniform coverage is obtained by overlapping the strips by half the width of a strip. Suppose, for example, the strip-profiling involves four parallel receiver lines straddling each shot line. Then, to obtain a square grid of datum points, the receiver line spacing could equal the in-line receiver group interval, and shot lines could be separated by twice the group interval.

3-D surveys generated from either parallel straight lines or parallel overlapping strips can provide all the information required to perform 3-D migration and to solve for interval velocity in an arbitrarily layered model having curved interfaces and isotropic, locally homogeneous velocity layers. However, as mentioned above, all the moveout information observed pertains to essentially one direction, the direction of shooting. If the azimuths between sources and receivers were distributed more evenly, we could obtain moveout information directly in three or more directions and thus have the luxury of redundancy for improved stability of velocity estimates.

One way to obtain a variety of azimuths between source and receiver while maintaining a regular grid of datum points is to have shotpoints and receiver stations located uniformly on separate straight lines that intersect at right angles. A reflection trace for every shot on the one line will be recorded at each receiver station on the other line. Such a *crossed array* can be viewed as a basic element for a 3-D seismic survey. It yields single-fold coverage over a rectangular area of the subsurface that is half the length of the line of shots by half the length of the line of receivers. The total number of datum points is $N_S N_R$, where N_S and N_R are the number of shotpoints and number of receiver stations, respectively, in the crossed array.

Appropriate translation of the crossed array provides multiple areal CDP coverage of the subsurface. Suppose the translation is in just one direction, the direction (x-direction) along the line of receivers. Depending upon the increment of translation distance, we obtain, say, N_1-fold CDP multiplicity along N_S parallel lines. Along the receiver line, the data are no different from that of a conventional, straight 2-D line. Azimuths between sources and receivers vary in the lines that are removed from the line of receivers. For any given CDP, however, the component of vectors from sources to receivers normal to the translation direction remains constant. Therefore, the zero offset time t_0 together with the three coefficients of the quadratic form [see formula (6.13)] describing the normal moveout time cannot be determined completely, because all the 4×4 determinants of the linear equations involved become zero, regardless of the degree of coverage in the x-direction. However, if the translation is also carried out in the direction of the shots, i.e., the y-direction, then the multiplicity in this direction is N_2-fold and the total multiplicity is $N_1 N_2$. Then the respective linear equations have an unambiguous solution if $N_1 = N_2 = 2$. They are overdetermined if N_1 or N_2 or both of them are integers larger than 2. Knowing the coefficients of the quadratic form of the normal moveout times, we can obtain estimates of the CDP-traveltime curves for all directions and each CDP point.

In order to get satisfactory velocity information from normal moveout times, it is highly desirable to have both N_1 and N_2 exceed 2 essentially. This may become prohibitive because of economical reasons. However, by combining the results of neighboring common datum points, this difficulty can mostly be bypassed.

Increasing the multiplicity of coverage in more than one direction also helps solve the critical problem of estimating statics for 3-D data.

We should reemphasize here that overlapped strip profiles in just one direction will provide sufficient information for the determination of interval velocity in the presence of 3-D structure (see section 9.1.3). Confining the survey to just one direction merely restricts the desirable redundancy that translation or roll along techniques in the x- and y-directions or linear surveying in three directions would provide. On the other hand, it simplifies the problem of dynamic corrections, especially if the selected direction coincides with that of the prevailing strike, if any exists.

We will not go into further detail here on the broad variety of possible field configurations for areal seismic data acquisition. They can all be derived essentially by local transformations using the simple principles given here.

5.2.3 Summary

CDP gatherings in line and in areal surveys consist of collections of traces having common midpoints between shotpoint and receiver station. In some cases, the strict CDP is replaced by a multiplicity of nearly common datum points. For line surveys, the vectors leading from shotpoint to receiver station maintain a common direction. In areal surveys, these vectors may have a variety of directions. A modified type of line shooting, called strip shooting, allows determination of the total gradient of the zero-offset reflection time. Genuine 3-D seismic surveys can further provide CDP traveltime curves in all directions.

6 CDP stacking and NMO velocity

So long as layer interfaces in the earth's subsurface have only moderate curvature, *CDP reflections* for primaries and symmetric multiples fall upon (symmetric) *CDP reflection time curves* that, for small offsets, are approximately hyperbolic. In contrast, the CDP reflections for asymmetric multiples need not have their apexes at zero offset (Levin and Shah, 1977).

It is difficult to conclude from one CDP reflection time curve alone whether the geologic interfaces are moderately curved or plane-dipping reflectors. So long as lateral velocity changes within layers can be excluded or presumed known the fact that interfaces are dipping or curved can be deduced from the normal reflection time curves of one seismic profile or from neighboring CDP gathers. The more the subsurface is approximated by a *homogeneous plane-layer model,* the better the CDP reflection time curves approximate hyperbolas.

For CDP reflections from reflectors that are not too shallow, the dynamic characteristics (i.e., amplitudes and wave shapes) generally change little with shot-geophone offset. At earlier times, where critical angles and large deviations from hyperbolas occur, these characteristics are more variable.

CDP stacking, also known as *horizontal stacking,* involves summing primaries along their interpreted primary CDP reflection time curves. So long as layers are nearly planar and horizontal, and subsurface velocities tend to increase with depth, CDP reflection time curves for multiples will be more curved than those of primaries having the same two-way zero-offset times.

The CDP stacking process can be studied from both a signal- and ray-theoretical point of view. In computing interval velocities and solving the inverse traveltime problem, we draw upon this latter aspect of the process. Since it involves both signal enhancement and noise suppression, CDP stacking also can be described in communication theoretical terms. It can, in fact, be looked upon as a multichannel filter operation that transforms traces of a CDP gather into an (*optimum*) *stacked trace.* Such a stacked trace is expected to approximate a noise- and multiple-free trace that would result from a hypothetical, coinciding source-receiver pair at the CDP location.

The degree to which an actual stacked trace approximates the desired result depends upon various parameters such as the number of traces, spread-length, and complexity of subsurface geology. Nevertheless, the CDP stacking process generally yields significant improvements in the ratio of signal to random noise and primary to multiple reflections.

The most common method of approximating the desired stacked trace involves summing all primary CDP reflections into the apexes of the hyperbolas that best approximate the actual primary CDP reflection time curves. The primaries in an optimum CDP stacked trace then get enhanced against a background of random noise, and the multiples tend to be suppressed as their ensemble of time-distance curves normally conflicts with the primary curve. The signal or communication theoretical aspects of the CDP stacking process are dealt with in a number of papers (Meyerhoff, 1966; Cressman, 1968; Galbraith and Wiggins, 1968; Buchholtz, 1972; Dunkin and Levin, 1973; White, 1977; Bading and Krey, 1976).

In the following, we discuss only those ray theoretical aspects useful in computing interval velocities and solving the inverse traveltime problem.

6.1 Ray theoretical definitions

So long as layer boundaries and layer velocities are analytic, the traveltime of a ray and the ray intersection point at the surface of the earth are analytical functions of the ray starting point and the starting direction of the ray. Moreover, a primary (or symmetric multiple) CDP reflection time curve always constitutes a (half-sided) curve symmetric about zero-offset. Therefore, regardless of the assumed model, reflection time can be expanded into a Taylor series of the following form

$$t(r) = t(0) + a_2 r^2 + a_4 r^4 + a_6 r^6 + \dots . \tag{6.1}$$

The variable r designates the shot-receiver distance in a CDP profile. The problem of finding the radius of convergence of equation (6.1) will not be discussed.

The function $\Delta t(r) = t(r) - t(0)$ is called *normal moveout* (NMO) or delta-T. It is frequently evaluated for the maximum offset $r = r_{\text{MAX}}$ of a CDP profile or for a specified unit distance such as 1 km. The difference in NMO for a primary and multiple CDP reflection arriving at the same zero-offset time is a good measure for how well a multiple can be suppressed in the CDP stacking process. An alternative series expansion is obtained by squaring equation (6.1) (see chapter 4):

$$t^2(r) = t^2(0) + r^2 / V_{\text{NMO}}^2 + b_4 r^4 + b_6 r^6 + \dots . \tag{6.2}$$

This series offers the advantage of including hyperbolas which must result for homogeneous plane (single) layer models. For plane horizontal isovelocity layers, V_{NMO} in equation (6.2) equals the RMS velocity V_{RMS}. For more general models, however, V_{NMO} differs from V_{RMS}; V_{NMO} denotes *normal moveout velocity*. The coefficient a_2 in equation (6.1) can be expressed in terms of V_{NMO} and $t(0)$ as $a_2 = 1/[2t(0)\,V_{\text{NMO}}^2]$.

Numerous computational ray-tracing experiments (Robinson, 1970a; Ursin, 1977) have verified that, for horizontal layer models, the *small-spread hyperbola* $t_H^2(r) = t^2(0) + r^2/V_{\text{NMO}}^2$ as given by the first two terms in equation (6.2) is a better approximation for the reflection time curve $t(r)$ than is the *small-spread parabola* $t_p(r) = t(0) + r^2/[2t(0)\,V_{\text{NMO}}^2]$ given by the first two terms in equation (6.1). For reasons of continuity, the superiority of the hyperbolic assumption holds also for more complex subsurface models. Therefore, in this monograph, we will always prefer the small-spread hyperbola to the small-spread parabola.

CDP rays for finite offset r generally do not reflect at the *normal incidence point* (NIP) (Figure 5-2). For second-order (i.e., for small offset) approximations

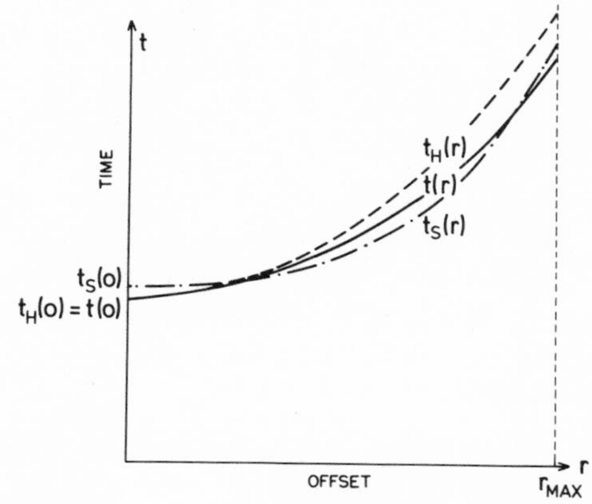

FIG. 6-1. $t(r)$ = CDP reflection time curve; $t_s(r)$ = best fit stacking hyperbola; $t_H(r)$ = small-spread hyperbola.

to CDP reflection time curves, however, one can assume that all CDP rays pass through the NIP. The CDP ray family then agrees sufficiently well with a ray family belonging to a hypothetical wave propagating away from a point source at the NIP. The proof for this statement as given previously by Krey (1976) is reviewed in Appendix D.

Later we will express V_{NMO} analytically in terms of seismic parameters along the normal incidence ray by using wavefront curvature laws. Expressing a_i and $b_i (i = 4, 6, 8, \ldots)$ in terms of parameters along the normal ray is also possible, in principle, but is not considered here. A solution to this particular problem for the plane, horizontally layered case was provided in section 4.1.

The hyperbola-like CDP reflection time curves can usually be well approximated, up to some maximum offset r_{MAX}, by the following hyperbola (Figure 6-1)

$$t_s^2(r) = t_s^2(0) + r^2 / V_s^2. \tag{6.3}$$

Expression (6.3) can be conceived as resulting from best fitting a hyperbola over a specified portion of the actual CDP reflection time curve.

V_s is the *(optimum) stacking velocity* and $t_s(0)$ is the *(optimum) stack time.* Both parameters in equation (6.3) generally depend on maximum offset r_{MAX} as well as on the profile azimuth. This is certainly an unwelcome property for the seismic interpreter because he would prefer to associate any determined velocity or traveltime with the geology only, uninfluenced by the geometry of the spread and profile azimuth. V_s is, in fact, a very complicated function when expressed analytically in ray-theoretical terms. Practically, however, from a signal theoretical or processing point of view, it can be easily determined and explained by means of *stacking-velocity analysis* as described in chapter 10.

One way of implementing stacking-velocity analysis (SVA) involves summing traces in a CDP family along a number of *test hyperbolas* or *analysis hyperbolas* for specified zero-offset, two-way reflection times. Each test hyperbola is characterized by the *test stacking velocity* used. In such a velocity analysis, the absolute value of the sum is a measure of the coherency of trace amplitudes. A hyperbola for which coherence is locally a maximum may characterize the NMO of a primary reflection. Equation (6.3) may represent such an optimum stacking hyperbola.

Optimum stacking hyperbolas must be determined for all primary CDP reflections in a CDP gather in order to obtain an optimum CDP-stacked trace. It is obvious that as the spread-length decreases (i.e., $r_{\mathrm{MAX}} \rightarrow 0$), $V_s \rightarrow V_{\mathrm{NMO}}$ and $t_s(0) \rightarrow t(0)$. Though still somewhat complicated to describe analytically, V_{NMO} is considerably simpler and more valuable than V_s for the purpose of computing interval velocities.

In general, the difference $B = V_s - V_{\mathrm{NMO}}$, called the *spread-length bias,* depends on both r_{MAX} and profile azimuth. Since it is identically zero for a homogeneous plane layer model, irrespective of r_{MAX}, it provides some indication about the *heterogeneity* of the velocity layers above a considered reflector. The smaller the value for B, the better the system of layers approximates a homogeneous plane layer model. The spread-length bias can be expressed analytically (Al-Chalabi, 1973) for a plane horizontally layered earth model, if one considers equation (6.3) as resulting from a least-squares fit to a true CDP reflection time curve. Let us consider the function $\delta t(r) = t(r) - t_s(r)$. It generally assumes very small values for typically used values of r_{MAX}. To ensure that all primary CDP reflections stack sufficiently well in-phase, the difference must not exceed some upper limit. As $\delta t(r)$ generally increases with increasing offset r_{MAX}, the potential for degradation of stacking signals can counteract the other advantages in resolution gained by increasing the spread-length. To ensure that, for long seismic spreads, the CDP stacking process does no more harm than good to the traces of a CDP gather, one occasionally stacks near and far traces independently. This special effort, in effect, provides an approximation to higher-order terms in r. It enables one to increase and ascertain the accuracy of a stack.

Plotting only traces of a fixed source-receiver offset for the sequence of CDPs results in a *common-offset section.* Plotting optimum CDP-stacked traces for the sequence of CDPs results in a *CDP-stacked (time) section.* When data are properly stacked, the stacked primary reflection times closely approximate two-way normal times and thus can be simulated with the help of normal-incidence ray modeling (Taner et al, 1970).

Plotting two-way normal times for a selected reflector as a function of the coordinates on the earth's surface results in the *(two-way) normal reflection time function.* This function may be displayed as a *time section* or *time map* (isochron map). Even for smoothly curved subsurface reflectors and constant layer velocities, this function need not be single valued; it can have cusps and singular points associated, for example, with buried focuses. Normal reflection time maps can be looked upon as representing the traveltimes of hypothetical waves that originate at time zero at the various reflectors and travel to the surface with half the actual local velocity of layers.

Time-to-depth migration, yet to be described, can be viewed as the inverse process by which such *hypothetical wavefronts* at the surface are propagated back down to those reflecting interfaces. Hypothetical wavefronts of various kinds will later be called upon to help solve different inverse traveltime problems.

The type of hypothetical wavefront just described will be referred to as a *normal wavefront.* The process of shrinking it back to a subsurface reflector through a known velocity field, in essence, transports surface measurements (i.e., two-way traveltimes) downward, changing them into respective subsurface parameters (i.e., depth maps). Whenever the geology is complex, it is most unlikely that normal rays will be confined to a *vertical seismic plane.* They then represent 3-D trajectories in 3-D space; hence, the inverse process of migration must take into account the full 3-dimensionality of the wave field.

6.2 Normal movement velocity

NMO velocity, expressed analytically in terms of seismic parameters along a normal incidence ray, provides an important key to solving inverse traveltime problems. The rest of this chapter is devoted to obtaining analytic expressions for V_{NMO} for selected subsurface models of increasing degrees of complexity. Practical problems related to removing the spread-length bias between V_S and V_{NMO} and performing other necessary corrections on observed surface measurements will be discussed in chapter 10.

For the time being, we shall assume that sufficiently short seismic spreads are available, thus allowing direct determination of the NMO velocity and two-way normal time for primary CDP reflections of a selected sequence of reflecting beds. The reader thus will better appreciate the "physical significance" of both quantities. He will not be confused by the various corrections that must be considered to obtain V_{NMO} from actual data. Nevertheless, we keep in mind that, since V_{NMO} relates in fact to an infinitely short CDP profile for which normal moveout is negligible, NMO velocity has to be considered strictly as a mathematical abstraction.

6.2.1 Plane horizontal layers

In sections 4.1 and 4.3, we derived the following small-spread hyperbola for the primary CDP reflection time curve of a reflecting interface

$$t_H^2(r) = t^2(0) + r^2 / V_{\mathrm{NMO}}^2 . \tag{6.4}$$

Here, the NMO velocity V_{NMO} is the same as the RMS velocity V_{RMS} as defined in equation (4.18). The wavefront curvature laws provide the simplest and most appealing way of establishing formula (6.4). The formula, originally obtained by Dix (1955), is included in a more general expression for vertically inhomogeneous layers, previously discussed by Krey (1951, 1954).

For all values of r, the traveltimes of a small-spread hyperbola for reflections beneath the first one exceed actual CDP reflection times. Stated differently, $V_s \geqq V_{\mathrm{NMO}}$, where the equality holds only for the reflector beneath the first layer. It can also be shown that $t_s(0) \geqq t(0)$. For a given r_{MAX}, the difference between $t_s(0)$ and $t(0)$ will generally be much smaller than the difference between V_S and V_{NMO}. Al-Chalabi (1973) and Bortfeld (1973) have proven that the two inequalities are valid for all values of r_{MAX}. These inequalities do not necessarily hold true in the presence of curved and dipping velocity boundaries.

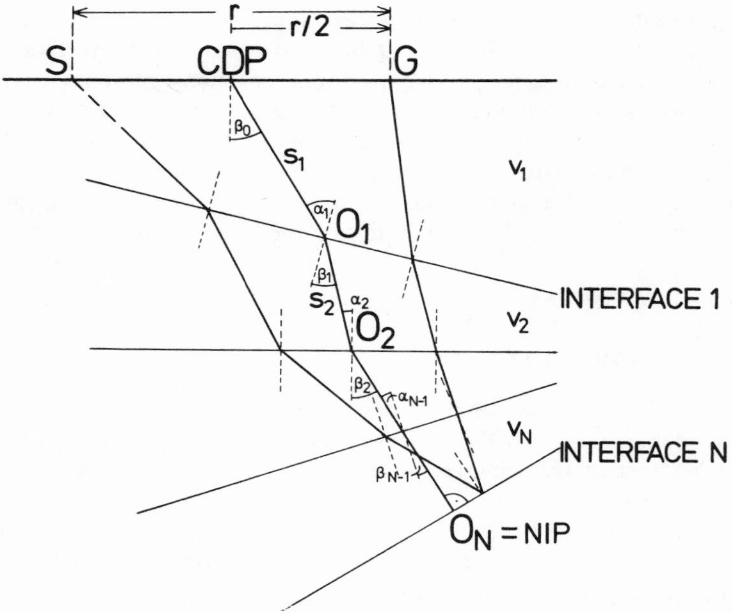

FIG. 6-2. 2-D plane-dipping layer model featuring a normal ray and a CDP ray for offset *r*.

6.2.2 2-D plane-dipping layers

Figure 6-2 shows a 2-D plane-dipping, isovelocity-layer model. A normal ray is traced from CDP to the *N*th reflector. Indicated are the angles of incidence and refraction α_i and β_i ($i = 1, \ldots, N - 1$), as well as the ray emergence angle β_0. A CDP ray for offset r is also traced. For small values of r, the traveltime along the reflected CDP ray between source S and receiver G is well-approximated by the sum of the times from S to the NIP and from the NIP to G (Appendix D). With equation (4.41), one can approximate the time from S to the NIP as

$$t_{NS} = t(0)/2 \pm \frac{\sin \beta_0}{v_1} \hat{x} + \frac{\cos^2 \beta_0}{2\, v_1 R_0} \hat{x}^2 + \ldots, \tag{6.5}$$

and from the NIP to G as

$$t_{NG} = t(0)/2 \mp \frac{\sin \beta_0}{v_1} \hat{x} + \frac{\cos^2 \beta_0}{2\, v_1 R_0} \hat{x}^2 + \ldots. \tag{6.6}$$

The $\pm$ sign in (6.5) and the $\mp$ sign in (6.6) allow for an interchange of point S with G. As β_0 is always positive, the upper sign in both formulas has to be chosen in a configuration as given in Figure 6.2 where the $\hat{x}$-axis points from right to left.

The quantity $t(0)$ is the two-way normal time of the normally reflected wave for a coincident source-receiver pair at the CDP. R_0 is the radius of curvature of a hypothetical wave originating at the NIP and emerging at the CDP. The hypothetical wavefront is subsequently referred to as the NIP wavefront. The small-spread parabola approximating the CDP reflection time curve is obtained from equations (6.5) and (6.6) as

$$t_P(r) = t_{NS} + t_{NG} = t(0) + \frac{\cos^2 \beta_0}{v_1 R_0} \left(\frac{r}{2}\right)^2. \tag{6.7}$$

Squaring both sides of equation (6.7) and ignoring the term involving r^4 results in the small-spread hyperbola

$$t_H^2(r) = t^2(0) + \frac{t(0) \cos^2 \beta_0}{2 v_1 R_0} r^2, \tag{6.8}$$

from which one can conclude that

$$V_{\mathrm{NMO}}^2 = \frac{2 v_1 R_0}{t(0) \cos^2 \beta_0}. \tag{6.9}$$

From the development of formula (4.41), R_0 and β_0 need not necessarily pertain to an emerging ray confined to a vertical seismic plane as assumed in equation (6.9). Formula (6.9) is also valid with respect to any oblique ray emergence plane that includes the emerging normal ray and the seismic profile. In section 4.3, a scheme was described by which R_0 for the emerging hypothetical NIP wavefront can be expressed as a function of seismic parameters along the normal incidence ray, i.e.,

$$R_0 = \frac{1}{v_1} \sum_{j=1}^{N} v_j^2 \Delta t_j \prod_{k=1}^{j-1} \left(\frac{\cos^2 \alpha_k}{\cos^2 \beta_k}\right). \tag{6.10}$$

Here, Δt_j is the one-way time along the normal ray in the jth layer. An almost identical formula to equation (6.9) was derived by Shah (1973b) where, for the same model, V_{NMO} is expressed in terms of the radius of curvature of a reflected wave that originates and is recorded at the CDP. If $\bar{R}_0$ is the radius of a curvature of the reflected wavefront at CDP, then one can verify [see formula (4.55)] that $\bar{R}_0 = 2 R_0$. This simple result no longer holds true for more complex velocity models.

It is obvious that V_{NMO}, as discussed here, reduces to V_{RMS} for plane horizontal isovelocity layers. This behavior will also hold for the more complex expressions for V_{NMO} that we shall subsequently derive. These expressions have forms related to those discussed by Krey (1976), Hubral (1976c), Gol'din and Černjak (1976), and Nakashima (1977a).

6.2.3 2-D curved layers

Formulas (6.8) and (6.9) hold for any layered model so long as (1) V_1 is constant, (2) R_0 pertains to a wavefront originating at NIP, and (3) R_0 is measured in a plane defined by the emerging ray and profile. To express V_{NMO} for the curved isovelocity layer model of Figure 6-3, one must again derive R_0 for the emerging hypothetical NIP wavefront along the normal ray from the NIP

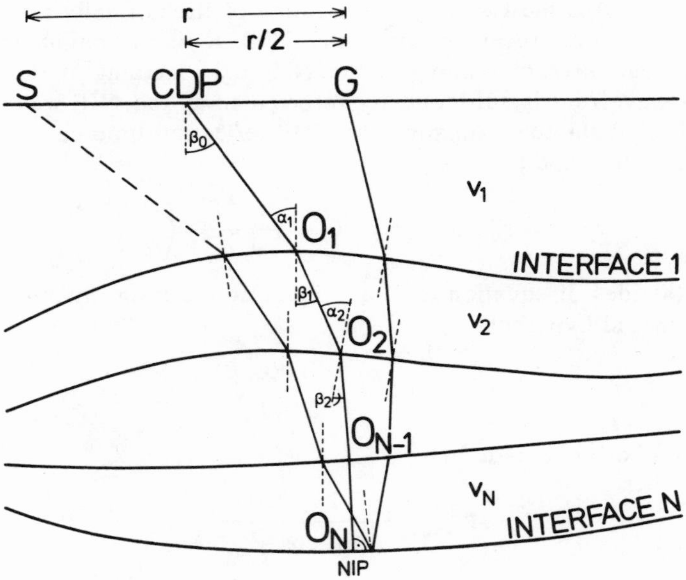

FIG. 6-3. 2-D curved layer model featuring a normal ray and a CDP ray for offset *r*.

to the CDP. Using the wavefront curvature laws (4.58) and (4.59), one can compute R_0 by a recursion as illustrated here for a three-layer case.

Radius of curvature on lower side of 0_2:

$$R_{I,2} = v_3 \Delta t_3 .$$

Radius of curvature on upper side of 0_2:

$$\frac{1}{R_{T,2}} = \frac{v_2 \cos^2 \beta_2}{v_3 \cos^2 \alpha_2} \frac{1}{R_{I,2}} + \frac{1}{\cos^2 \alpha_2} \rho_2 \frac{1}{R_{F,2}},$$

$$\rho_2 = \frac{v_2}{v_3} \cos \beta_2 - \cos \alpha_2 .$$

Radius of curvature on lower side of 0_1:

$$R_{I,1} = R_{T,2} + v_2 \Delta t_2 .$$

It is obvious how continuing the recursion will eventually yield R_0 for the emerging NIP wavefront. For the three-layer model, the following continued fraction results

$$R_0 = \frac{1}{v_1}\left[s_1 v_1 + \left(\frac{\rho_1}{v_1 \cos^2 \alpha_1} \frac{1}{R_{F,1}} + \left\{ s_2 v_2 \frac{\cos^2 \alpha_1}{\cos^2 \beta_1} + \left[\frac{\rho_2 \cos^2 \beta_1}{v_2 \cos^2 \alpha_1 \cos^2 \alpha_2} \frac{1}{R_{F,2}} + \left(s_3 v_3 \frac{\cos^2 \alpha_1 \cos^2 \alpha_2}{\cos^2 \beta_1 \cos^2 \beta_2} \right)^{-1} \right]^{-1} \right\}^{-1} \right)^{-1} \right]. \tag{6.11}$$

It should be remarked that in the presence of curved refracting velocity boundaries, R_0 can be negative; consequently, the moveout is negative, and V_{NMO} becomes an imaginary quantity. Negative moveout is not unlikely for reflections from complex structures.

6.2.4 3-D curved layers

Figure 6-4 shows a straight CDP profile in some arbitrary direction above a 3-D isovelocity layer model. A normal ray is traced from CDP to the Nth reflector. An $[x_s, y_s]$-coordinate system is placed at the CDP in the familiar way so that the positive x_s-axis points in the direction of the emerging normal ray projected into the surface of the earth. The $[x_s, y_s]$ system for a selected reflector will generally differ from the $[x_s, y_s]$ systems of other reflectors.

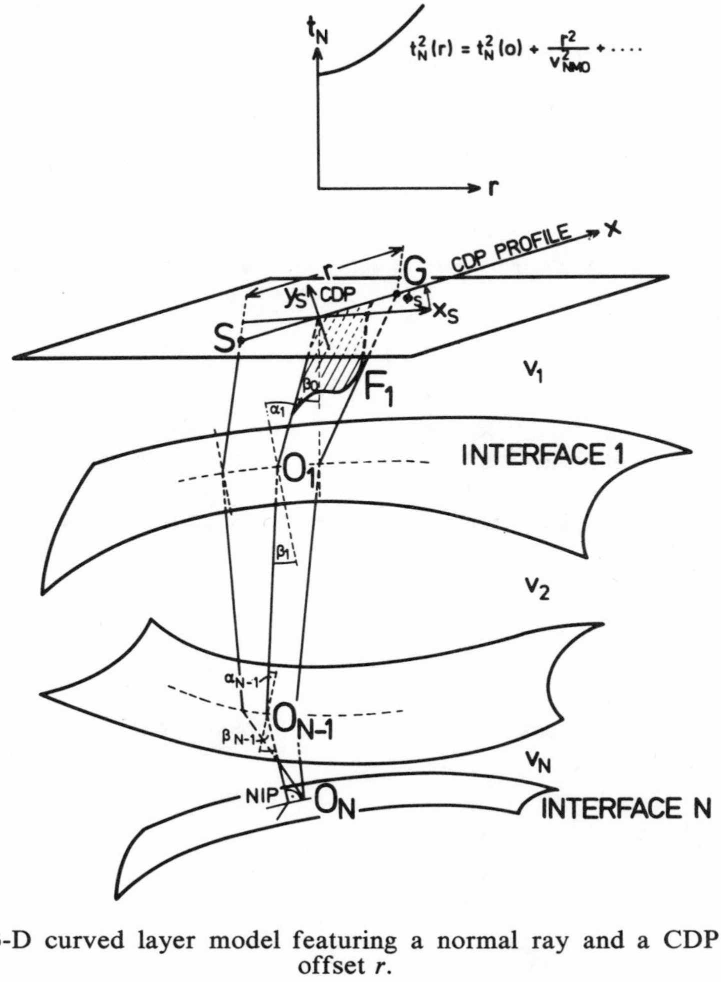

FIG. 6-4. 3-D curved layer model featuring a normal ray and a CDP ray for offset r.

The straight CDP profile in Figure 6-4 is oriented in the direction with azimuth ϕ_s with respect to the $[x_s, y_s]$ system associated with the Nth interface. In order to express V_{NMO} in a specified profile in terms of parameters along the normal ray, we place a hypothetical source at the NIP and compute γ_0 and R_0 for the emerging NIP wave at the CDP within plane F_1. The generally oblique plane F_1 is defined by the emerging normal ray and the line of profile. Taking into account equation (4.41), the quantity V_{NMO} is available from equation (6.9) as

$$V^2_{NMO} = \frac{2 v_1 R_0}{t(0,0) \cos^2 \gamma_0}. \tag{6.12}$$

Equation (6.12) can also be written with respect to the $[x_s, y_s]$ system in terms of the curvature matrix $\mathbf{A}_0$ of the hypothetical NIP wavefront emerging at the CDP. If $t(x_s, y_s)$ represents the reflection time from point S at $(-x_s, -y_s)$ to the NIP and back to point G at (x_s, y_s), then one obtains, for the hyperbolic second-order approximation of the CDP reflection time curve,

$$t^2(x_s, y_s) = t^2(0,0) + \frac{t(0,0)}{2 v_1} \mathbf{X}_s \mathbf{S}_0 \mathbf{A}_0 \mathbf{S}_0 \mathbf{X}_s^T;$$

where (6.13)

$$\mathbf{X}_s = (x_s, y_s) \quad ; \quad \mathbf{S}_0 = \begin{bmatrix} \cos \beta_0 & 0 \\ 0 & 1 \end{bmatrix}.$$

From equation (6.13), we conclude that the NMO velocity describes a conic section (typically an ellipse) when considered as a function of the azimuth ϕ_s, i.e.,

$$\frac{1}{V^2_{NMO}(\phi_s)} = \frac{t(0,0)}{2 v_1} \mathbf{e} \mathbf{S}_0 \mathbf{A}_0 \mathbf{S}_0 \mathbf{e}^T, \tag{6.14}$$

where $\mathbf{e} = (\cos \phi_s, \sin \phi_s)$. The principal axes of the conic section generally do not coincide with the directions of the x_s- and y_s-axes. Applying the wavefront curvature transmission and refraction laws (4.36) and (4.38) to the NIP wavefront from the NIP to the CDP along the normal incidence ray provides $\mathbf{A}_0$.

Let us perform the first few recursive steps from NIP to CDP. We will use the following notation

$$\rho_i = \frac{v_i}{v_{i+1}} \cos \beta_i - \cos \alpha_i;$$

$$\mathbf{S}_{I,i} = \begin{bmatrix} \cos \alpha_i & 0 \\ 0 & 1 \end{bmatrix} \quad ; \quad \mathbf{S}_i = \begin{bmatrix} \cos \alpha_i / \cos \beta_i & 0 \\ 0 & 1 \end{bmatrix},$$

and the subsequent law of matrix algebra:

$$(k \mathbf{F} \mathbf{H})^{-1} = \frac{1}{k} \mathbf{H}^{-1} \mathbf{F}^{-1}, \tag{6.15}$$

where k is a scalar constant and $\mathbf{F}$ and $\mathbf{H}$ are 2×2 matrices.

The first few recursive steps can now be expressed in a form which will lead to a closed-form expression for $\mathbf{A}_0$.

$$\mathbf{A}_{I,N-1}^{-1} = \frac{1}{v_N}(s_N v_N \mathbf{I}),$$

$$\mathbf{A}_{I,N-2}^{-1} = \frac{1}{v_{N-1}}\Bigg\{s_{N-1}v_{N-1}\mathbf{I} + \Bigg[\frac{\rho_{N-1}}{v_{N-1}}\mathbf{D}_{N-1}^{-1}\mathbf{S}_{I,N-1}^{-1}\mathbf{B}_{N-1}\mathbf{S}_{I,N-1}^{-1}\mathbf{D}_{N-1} + (s_N v_N \mathbf{D}_{N-1}^{-1}\mathbf{S}_{N-1}\mathbf{S}_{N-1}\mathbf{D}_{N-1})^{-1}\Bigg]^{-1}\Bigg\},$$

$$\mathbf{A}_{I,N-3}^{-1} = \frac{1}{v_{N-2}}\Bigg[s_{N-2}v_{N-2}\mathbf{I} + \Bigg(\frac{\rho_{N-2}}{v_{N-2}}\mathbf{D}_{N-2}^{-1}\mathbf{S}_{I,N-2}^{-1}\mathbf{B}_{N-2}\mathbf{S}_{I,N-2}^{-1}\mathbf{D}_{N-2} + \Bigg\{s_{N-1}v_{N-1}\mathbf{D}_{N-2}^{-1}\mathbf{S}_{N-2}\mathbf{S}_{N-2}\mathbf{D}_{N-2} + \Bigg[\frac{\rho_{N-1}}{v_{N-1}}\mathbf{D}_{N-2}^{-1}\mathbf{S}_{N-2}^{-1}\cdot \mathbf{D}_{N-1}^{-1}\mathbf{S}_{I,N-1}^{-1}\mathbf{B}_{N-1}\mathbf{S}_{I,N-1}^{-1}\mathbf{D}_{N-1}\mathbf{S}_{N-2}^{-1}\mathbf{D}_{N-2} + (s_N v_N \mathbf{D}_{N-2}^{-1}\mathbf{S}_{N-2}\mathbf{D}_{N-1}^{-1}\mathbf{S}_{N-1}\mathbf{S}_{N-1}\mathbf{D}_{N-1}\mathbf{S}_{N-2}\mathbf{D}_{N-2})^{-1}\Bigg]^{-1}\Bigg\}^{-1}\Bigg)^{-1}\Bigg].$$

In the case of three layers one gets the closed expression for $\mathbf{A}_0$,

$$\mathbf{A}_0 = v_1\Bigg[s_1 v_1 \mathbf{I} + \Bigg(\frac{\rho_1}{v_1}\mathbf{D}_1^{-1}\mathbf{S}_{I,1}^{-1}\mathbf{B}_1\mathbf{S}_{I,1}^{-1}\mathbf{D}_1 + \Bigg\{s_2 v_2 \mathbf{D}_1^{-1}\mathbf{S}_1\mathbf{S}_1\mathbf{D}_1 + \Bigg[\frac{\rho_2}{v_2}\mathbf{D}_1^{-1}\mathbf{S}_1^{-1}\mathbf{D}_2^{-1}\mathbf{S}_{I,2}^{-1}\mathbf{B}_2\mathbf{S}_{I,2}^{-1}\mathbf{D}_2\mathbf{S}_1^{-1}\mathbf{D}_1 + (s_3 v_3 \mathbf{D}_1^{-1}\mathbf{S}_1\mathbf{D}_2^{-1}\mathbf{S}_2\mathbf{S}_2\mathbf{D}_2\mathbf{S}_1\mathbf{D}_1)^{-1}\Bigg]^{-1}\Bigg\}^{-1}\Bigg)^{-1}\Bigg]^{-1}. \qquad (6.16)$$

For plane interfaces of arbitrary dip and strike, formula (6.16) reduces to formula (4.63).

Solution of the inverse traveltime problem and computation of interval velocities from NMO velocities will involve shrinking the NIP wavefront back into its hypothetical source. As will be shown later, it is not necessary to have $\mathbf{A}_0$ or V_{NMO} expressed in the compact form of equation (6.16) in order to compute interval velocities in a straightforward way.

6.2.5 Multiples

So far, V_{NMO} has been expressed only in terms of seismic parameters along normal rays of *primary* CDP reflections. Figures 4-5e–4-5f show representative rays for symmetric and asymmetric multiple CDP reflections. For symmetric multiple raypaths, the sequence of layers at which reflections occur is symmetrical about a central reflection. If, in addition, $r = 0$, the raypath shrinks to a single ray which is used twice; i.e., first for the downward path from the CDP to the NIP and then for the upward path from the NIP to the CDP. The procedure for computing wavefront curvatures and NMO velocities for such symmetric multiple paths does not differ from that for primary paths except that the reflection law is now included in the recursion.

Again one can place a hypothetical source at the NIP of the *multiple normal ray* and associate CDP rays with a multiple NIP wavefront. The NMO velocity of a symmetric multiple provides information about the subsurface similar to that obtained from the V_{NMO} of a primary reflection. Asymmetric multiples, however, cannot be associated with a hypothetical NIP wavefront; hence the curvature laws, however, cannot be applied in the same manner.

Though later inverse algorithms for computing interval velocities will only exploit information provided by primary CDP reflections only, it is quite conceivable that the redundant information in symmetric multiples could be considered just as well. The redundant information could be used either for obtaining inverse solutions or for reducing uncertainties in traveltime measurements related to primaries.

6.2.6 Summary

The NMO velocity of either a primary or symmetric multiple CDP reflection time curve can be expressed analytically in terms of seismic parameters along the normal ray. For small offsets, a CDP ray family for a particular reflector can then be associated with rays belonging to a hypothetical NIP wavefront, that originates at the normal incidence point of the normal ray. The wavefront curvatures of the NIP wavefront can be constructed recursively along the normal ray from bottom to top. This operation can be identified with a hypothetical forward wave propagation. V_{NMO} itself relates to some component of curvature of the emerging NIP wavefront. Note that V_{NMO} does not include the curvature of the reflector to which it belongs. The deviation of the actual CDP reflection time curve from a hyperbola has nothing to do with either the lateral displacement of the NIP or the spread of reflection points of a CDP family of rays across a reflector. This point can be easily demonstrated for homogeneous plane dipping layer models, which always provide hyperbolas as CDP reflection time curves.

Over a dipping layered earth, V_{NMO} varies with azimuth. The dependence is that of a conic section, generally an ellipse. The principal axes of the conic section normally coincide with the dip and strike directions of the considered reflector (in either time or depth) only when interfaces are planar and have a common strike direction. Moreover, in general, they do not relate to the direction of the emerging normal ray.

Solving the inverse traveltime problem can be viewed in terms of having an NIP wavefront shrink back into its hypothetical source. The computation of interval velocities from NMO velocities may consequently be viewed as a *downward continuation process.* Before we go into more details on this in chapter 9, we consider next the *time-migration process* which, in many respects, is complementary to that of CDP stacking.

The computation of V_{NMO} can be extended naturally to earth models with continuously changing layer velocities. The reader will find some information on this in Appendix F and in Hubral (1979a, 1979c).

7 Time migration

Exploration seismologists base their interpretation, skill, and judgment largely on the study of primary reflections for selected key horizons on time sections. These horizons are normally available in either CDP-stacked (i.e., unmigrated form) or *time-migrated form*. In regions of complex structure, the two forms of reflection can be markedly different; that is, at a common surface location the associated traveltimes to a particular reflector can differ considerably. Properly time-migrated sections can be expected to provide more realistic pictures of the geology—pictures which may not be simply inferred from their unmigrated counterparts, the CDP-stacked sections. The underlying principles of time migration are well described in the literature. We will, therefore, discuss only those ray-theoretical aspects that will assist later in better understanding problems related to computing interval velocities from *migration velocities*.

Various time migration schemes are used nowadays to transform either a CDP stacked section or common-offset sections obtained along a seismic line into a *time-migrated section*. One basic approach (known variously as Kirchhoff migration, Huygens-Fresnel migration, and diffraction migration) involves summing scattered signals (in a weighted manner) into the apex of diffraction curves. This approach is founded on the Kirchhoff integral solution to the wave equation. It can be implemented in various ways, all of which lead to almost identical results. In the presence of strong lateral velocity gradients, however, such schemes fail to position reflections correctly (sometimes grossly so) and in that sense fail to serve the desired function of migration. The terminology originated, however, at a time when with certain exceptions (e.g., Krey, 1951) analytical mirgration processes modeled the earth predominantly by a vertically inhomogeneous medium. It is only for such a medium that a complete mirgration can actually be obtained by any of these schemes. We shall refer to all such integral approaches as examples of *time migration*.

Time migration may be viewed as a particular type of image reconstruction that transforms the seismic record into an image of the earth's subsurface. CDP stacking is another type of image reconstruction. It is based on specular reflection assumptions and, therefore, works best for flat or plane, conformably dipping horizons. The image quality of a CDP stack degrades progressively as the geology becomes more complex, especially when the seismic wave field consists predominantly of diffracted energy. Often oil tends to accumulate in regions where geologic structure is anomalous and anything but horizontally stratified. In such regions, a CDP stack may not produce a sufficiently accurate or even usable picture of the subsurface. Time migration usually provides a less distorted image of the earth.

Integral approaches to time migration are similar in many respects to the process of CDP stacking and, like CDP stacking, have strong communication-

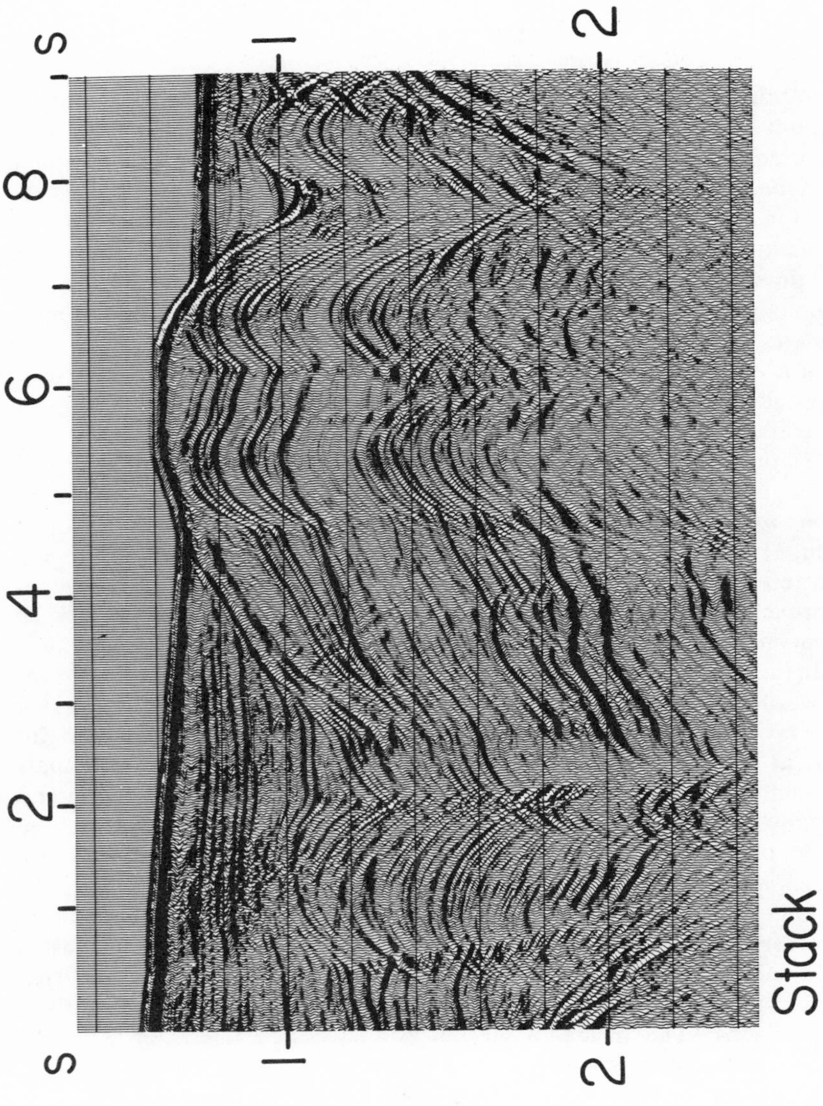

FIG. 7-1. CDP-stacked section (courtesy Western Geophysical Co.).

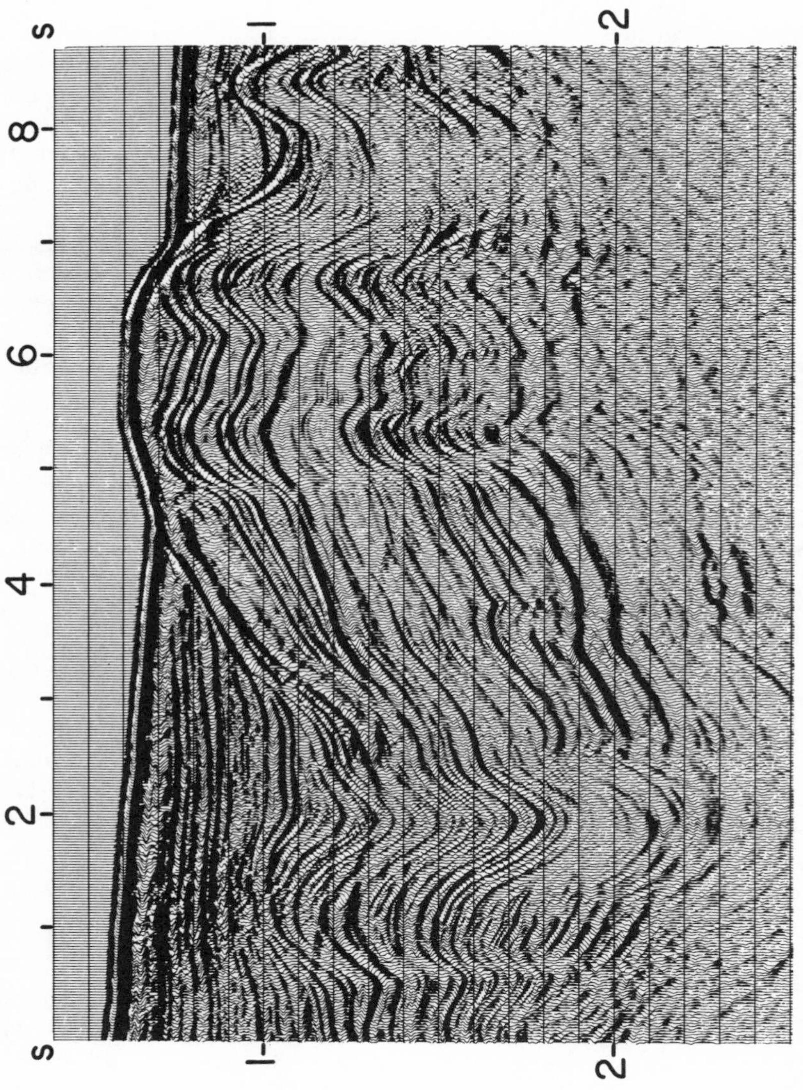

FIG. 7-2. Time-migrated section (courtesy Western Geophysical Co.).

theoretical (Rockwell, 1971; Sattlegger and Stiller, 1973; French, 1974, 1975; Newman, 1975; Hosken, 1978; Berkhout and Van Wulfften Palthe, 1979) as well as ray-theoretical (Hagedoorn, 1954; Hubral, 1977) aspects. Since these schemes transform one wave field into another, they must have their basis in the wave equation (French, 1975; Schneider, 1978) just as does *finite-difference* (so-called "wave equation") *migration* (Claerbout, 1971, 1976; Claerbout and Doherty, 1972; Loewenthal, Lu, Roberson, and Sherwood, 1976). The finite-difference approach, in fact, is just another particular scheme to implement time migration and no more deserves to be categorized as "wave equation" migration than do the integral approaches. Comparative studies of wave equation migration based on integral and finite-difference methods (Larner and Hatton, 1976) show that both often yield similar results. Because each is based on a different set of simplifying approximations, under certain circumstances one approach may be preferred over the other.

Time migration schemes aim essentially at focusing CDP-stacked or unstacked seismic reflections into a data form more suitable for interpretation. Like CDP-stacked data, time-migrated data are presented as a function of two-way time. In areas of complex geology, where wave fields have complexity in three dimensions, time-migrated data are most meaningful if obtained with 3-D time-migration schemes using 3-D recorded seismic data.

On time-migrated sections, diffracted energy is contracted, and weak, segmented reflections often appear in geologically more reasonable form than on CDP-stacked sections (e.g., the segmented events may appear more continuous or, in contrast, more clearly faulted). Faults and various structures frequently show up more distinctly. For comparison between a CDP stacked and time migrated section, see Figures 7-1 and 7-2.

Readers who would like to familiarize themselves with some aspects of time migration processes will find sufficient information in the above mentioned publications. We shall confine our attention to the ray theory since it is sufficient to investigate the relationship between traveltimes of reflections on CDP-stacked (or unmigrated) sections and time-migrated sections. Ray theory can also be used to reconstruct a depth model from the time-migrated section with the help of *migration velocities*. Our emphasis is with the 3-D case; the 2-D case is then self-explanatory.

7.1 Time migration versus time-to-depth migration

Figure 7-3 shows the general subsurface model that we shall assume. Using an elementary theory of time migration, we consider any subsurface feature and, in particular, a reflecting interface to consist of a *continuum of points* which, due to Huygen's principle, represent *scattering centers* for seismic waves. In this context, reflections arising at interfaces can be considered as the sum of the energy scattered by the continuum of interface points. Time-migration schemes, in essence, represent a search for all scattering centers. They *image* each subsurface point by detecting its scattered field at the earth surface.

Scattered fields are often observed on CDP-stacked sections as hyperboloid-like

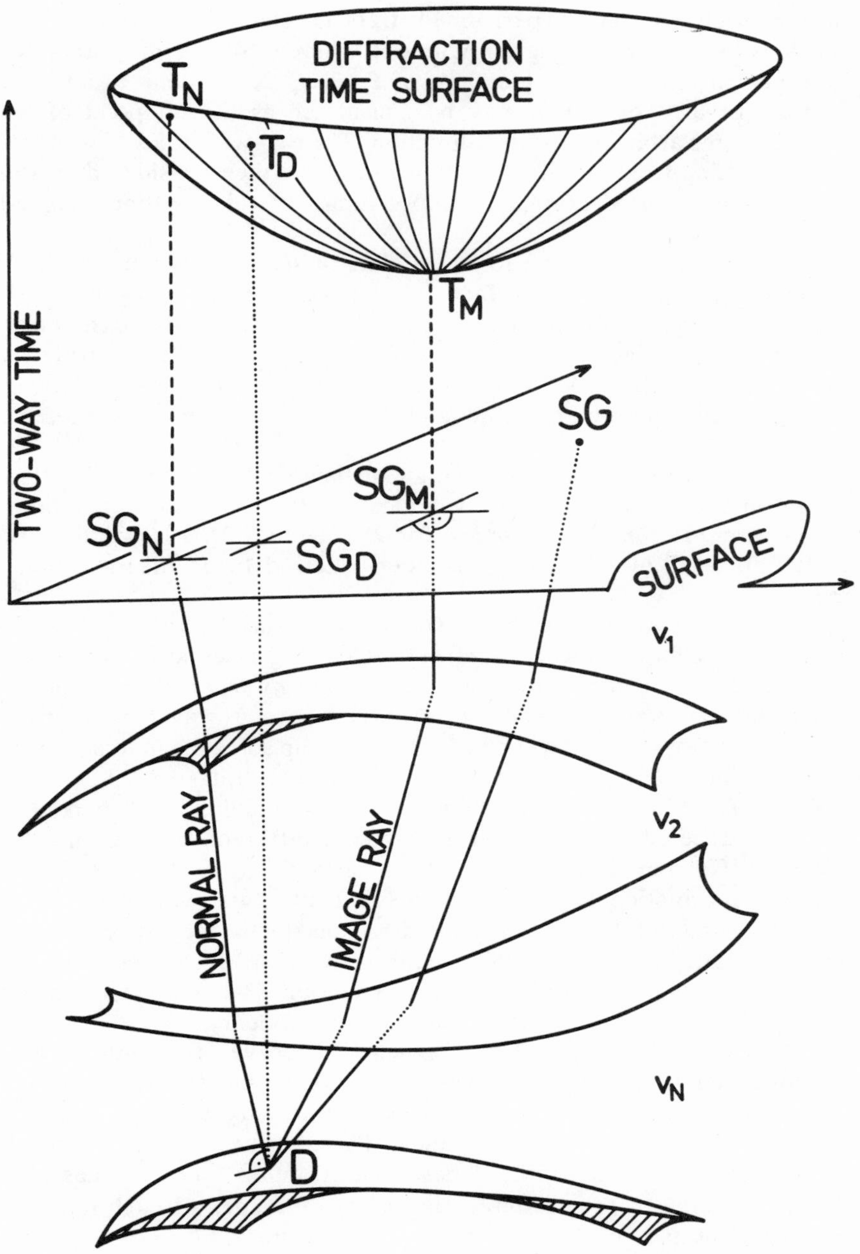

FIG. 7-3. 3-D curved layer model featuring a normal ray, image ray, and diffraction time surface for point scatterer D.

diffractions from reflector discontinuities or faults. More often, however, they cannot be detected by the eye because of complicated *interference* (Gardner et al, 1974b) with other portions of the wave field.

Time migration is most often performed after CDP stack. In such circumstances, one imagines a seismic recording configuration wherein coincident source-receiver pairs are placed at all surface locations of Figure 7-3. *SG* denotes an arbitrary source-receiver location which may be considered as the midpoint of a CDP profile along some arbitrary azimuth direction. The two-way times of well-stacked primaries resulting from multifold CDP coverage are then considered as approximations to the primary, normal reflection times of this fictitious single-fold, zero-offset profile.

The time required by a wave to travel along a scatter ray from a coincident shot-receiver pair *SG* on the surface to a subsurface scattering point *D* and back is of fundamental importance in the theory of time migration. Point *SG* is connected by a refracted ray with a scattering point *D* on the *N*th reflector. This ray obeys Snell's law between these two points. At each coinciding shot-receiver pair, one records some contribution from the pulse scattered back from *D*.

Plotting the two-way traveltime from all surface points *SG* to *D* as a function of the coordinates of the shot-receiver location results in the *diffraction time surface* or *migration moveout surface* for *D* (Figure 7-3). Elsewhere (French, 1975; Hubral, 1977), this surface has been referred to as the *reflection time surface.*

The time values that describe the diffraction time surface are twice the values associated with a hypothetical wavefront which originates at *D* and arrives at the surface. This wavefront is referred to as the *D-wavefront.* It is a hypothetical wavefront just as was the NIP wavefront or normal wavefront considered previously. The *D*-wavefront will be used later to help solve the inverse traveltime problem for time-migrated reflections in a way that will resemble the use of the NIP wavefront in inverse traveltime problems involving CDP-stacked data.

Each subsurface scattering center generates a different *D*-wave and, hence, a different diffraction time surface. These surfaces are well known (French, 1975) to be exact rotational hyperboloids only if centers are overlain by a medium of constant velocity. Likewise, only for a homogeneous acoustic medium does the Kirchhoff integral (Trorey, 1970; Hilterman, 1970, 1975; Schneider, 1978) provide an exact solution to the scalar wave equation. Only then is it appropriate to consider *Kirchhoff migration* for exact inverse modeling.

When rays refract because of the presence of curved first-order interfaces or inhomogeneous velocity layers, then diffraction time surfaces will deviate from rotational hyperboloids. Nevertheless, in time migration we approximate diffraction time surfaces by hyperboloids. The scheme involves a search for the hyperboloids that can be fitted best to diffraction time surfaces prior to summing the energy of the scattered signals into the apex of each hyperboloid time surface. The parameters of the *(optimum) migration hyperboloid* that can best be fitted to a major part of the diffraction time surface depend upon the width of the *aperture* (i.e., the number of traces used for obtaining each time-migrated output trace). Thus, the aperture half-width plays a role for the theory of time migration, equivalent to that played by the maximum offset in the theory of CDP stacking.

Certainly, where the geology is sufficiently complex, the migration hyperboloid will not adequately characterize the complicated shapes of diffraction time surfaces and, hence, those surfaces will not collapse satisfactorily about their scattering centers. Since the migration hyperboloid is the 3-D counterpart of the stacking hyperbola, this failure to migrate correctly occurs when the CDP stacking process becomes inapplicable for similar reasons.

In the next section, we discuss second-order approximations to diffraction time surfaces for 3-D layered media. These approximations are chosen in such a way that they are exact for infinitely small apertures in the vicinity of the apex of the diffraction time surface. As it turns out, the approximating second-order time surface in this limiting case is then again a hyperboloid, but not necessarily a rotational one. We call it the *small-aperture migration hyperboloid.* Its equivalent in the theory of (2-D) CDP stacking is the small-spread hyperbola. CDP stacking, in principle, could be extended to 3-D (with complexity) so that we would be dealing with a *small-spread hyperboloid.*

3-D time migration aims at summing all data values, i.e., the scattered energy upon each diffraction time surface, into a single output signal amplitude which is placed at the apex or minimum. The traveltime of the collapsed data thus coincides with the two-way minimum traveltime from D to the surface. By limiting the summation to a small portion of the diffraction time surface which contains the main energy contribution of D, one generally obtains good results. If the interface at D is predominantly a specular (nonrugged) reflector, most of the energy will be returned along the specular path (normal ray) and will thus be recorded in the region surrounding SG_N (see Figure 7-3).

The ensemble of rays connecting D with points on the surface of the earth includes two rays of special importance. One is the normal ray and the other is the ray emerging normally to the earth's surface, the ray we previously called the *image ray*. Because the layers in our model are assumed to be isotropic, the rays in Figure 7-3 are always normal to the hypothetical D-wavefront. Clearly, since the apex of the D-wave has a horizontal tangent, it coincides with the position of a vertically emerging ray; i.e., the image ray.

In time migration, signals scattered along the rays connecting D with all other surface points are time-migrated or repositioned at the apex point. Only the signal position related to an image ray remains the same before and after time migration. (Strictly speaking, only if the subsurface has a vertically inhomogeneous velocity distribution or plane dipping velocity boundaries can one be certain that diffraction time surfaces have only one apex. In arbitrary, laterally inhomogeneous media, they may have two or more apexes.)

The energy scattered from moderately curved or plane reflectors is dominated by a *specular* (reflected) energy return (French, 1974, 1975) along the rays normal to the reflectors. (The relative energy in the specular reflection increases with the frequency of the signal.) Detectors in the vicinity of SG_N will record the specular energy from the reflecting interface area around D. Thus, in order to image point D properly, we must ensure that SG_N falls within the *maximum aperture* range selected for the time-migration process.

The two-way traveltime from SG_N to D is the time at which the Nth reflecting horizon at D is observed on the primary normal reflection time map at SG_N. This particular time is usually well approximated by the stacked reflection time of a CDP gather with a common datum point at SG_N and arbitrary profile azimuth. In migration, the *primary reflection element* at time T_N is mapped (or moved, or "migrated") into an output element at time T_M (Figure 7-3).

Computation of a *normal reflection time map* for a given subsurface reflector, and overlying velocity distribution, involves tracing normal-incidence rays from points on the surface to points on the desired reflector (or vice versa). Likewise, computation of a *two-way image time map* for the same horizon requires the tracing of image rays from the surface. Image rays thus assume for the theory of time migration the same fundamental importance as do normal incidence rays for the theory of CDP stacking.

Image rays always emerge vertical at the surface; only in laterally homogeneous velocity media will they be vertical all the way to a reflector. In this case, time-migrated reflections fall vertically above their corresponding *depth elements*. Therefore, only in this case is time migration complete in the sense that subsequent time-to-depth conversion of time-migrated reflections can be achieved by merely vertically scaling the *two-way image time maps* to *depth maps*. No further *time-to-depth migration* (see sections 8.2 and 8.3) would be needed; conventional vertical scaling from time to depth would fully accomplish that task.

Because image rays refract at layer interfaces, we see that time migration cannot provide a *complete* migration for a reflector below a laterally inhomogeneous velocity medium. All the information in a time-migrated trace stems from subsurface points located along an image ray and not, as has frequently been believed, from points along the true vertical. Image rays, like normal-incidence rays, can cross one another in laterally inhomogeneous media. Where they do so, the particular subsurface points can have images at two or more points on a time-migrated section. This crossing corresponds to the fact that for a particular depth point, the diffraction time surface has more than one apex. Such a result is certainly not very appealing to a seismic interpreter but it is something he must live with unless he corrects time-migrated sections by one of the means described later.

Unlike normal-incidence rays, image rays hit their target reflector at *arbitrary* angles of incidence. Also unlike normal incidence rays, image rays from a given position on the earth's surface offer the unique advantage of being coincident (congruent) with one another irrespective of the reflector to which they are traced. As we shall see, this property of image rays can be exploited in various ways.

To collapse the diffractions in a time section successfully, one requires a less accurate subsurface velocity model than is required for accurate time-to-depth migration. This welcome property complements the classical accuracy requirements put on CDP stacking velocities and time-to-depth migration velocities. The latter need to be much more accurate than the former. Due to their inherent simplicity, image rays can be traced in an efficient effort to improve the velocity model. If computed image-ray intersections for a *continuous* time horizon fall, for instance, onto a *discontinuous* surface at depth, it is likely that the velocity field is incorrect. The velocity model can be altered and image rays traced until a plausible depth model results.

In laterally varying media, the necessary operation that must follow time migration is referred to as a *time-to-depth migration.* This operation not only can correct the lateral positions of events, it also can resolve the ambiguity of multiple images of a depth point.

T_D is a point (Figure 7-3) on the diffraction time surface vertically above D. It is related to a particular scatter ray from D to SG_D (not shown in the figure). If the energy of the scattered coherent signals along a diffraction time surface were summed into a signal at T_D, then the migration for the depth element D would be complete, and no further lateral displacement would be required in a subsequent conversion to depth. A search for the point T_D and the particular scatter ray from SG_D to the subsurface scatterer D, however, requires an accurate velocity model and extensive decision-making since the position of T_D within the (generally changing) diffraction time surface most likely will differ for different scatterers. Even if the energy were summed into T_D, the remaining conversion from time to depth would have to be based upon the particular ray from SG_D to D.

For practical reasons, in time migration the *image* of a subsurface scatterer is placed at the apex of its associated diffraction time surface; i.e., while time migration, as described above, thus honors *per definition* a practically feasible scheme, it is one which, in the presence of lateral inhomogeneities, does not generally provide a complete migration. The reflected element at time T_N recorded at SG_N does not migrate to T_D but to T_M, the apex of the diffraction time surface. This point is close to the apex of the actual optimum migration hyperboloid that, for the specified aperture surrounding T_M, provides the best approximation to the actual diffraction time surface.

Since image rays often can deviate significantly from the true vertical, time migrated primary reflections need not always provide a more truthful picture of subsurface reflectors than do primary stacked reflections. For deep reflectors falling below shallow curved or dipping velocity interfaces, horizontal displacements between D and SG_M (Figure 7-3) can be larger than those between D and SG_N. In this case, time migration may actually further distort a primary reflection normal time map rather than bring it closer to the position of the true reflector at depth. Note, however, that because of the congruency of image rays for all reflectors, such distortions are "global" as relative positions of nearby subsurface points will only be little distorted.[2]

In many circumstances, it is appropriate to look upon any time-migration scheme only as some matched time-varying filter process—one which improves unmigrated CDP-stacked reflections by contracting the (information concealing) diffractions into their apex locations.

Readers will note again that time migration discussed here in ray-theoretical terms is derived from the Kirchhoff integral representation of the wave equation for homogeneous media. We did not, however, limit the model to homogeneous media nor did we confine the discussion to vertically inhomogeneous media as used in the classical work of Hagedoorn (Hagedoorn, 1954; Walter and Peterson, 1976).

Those geophysicists introduced to the theory of "time migration" by the largely "ray-theoretical" approach given here cannot be surprised to find that the scheme does not always yield correct migrations. Those, on the other hand, introduced to the theory of time migration using the Kirchhoff integral or

[2] Thus, although the apparent positions of time-migrated reflections may be in error, the time-migrated section may, nevertheless, be more plausible and more easily interpreted then the original CDP stack.

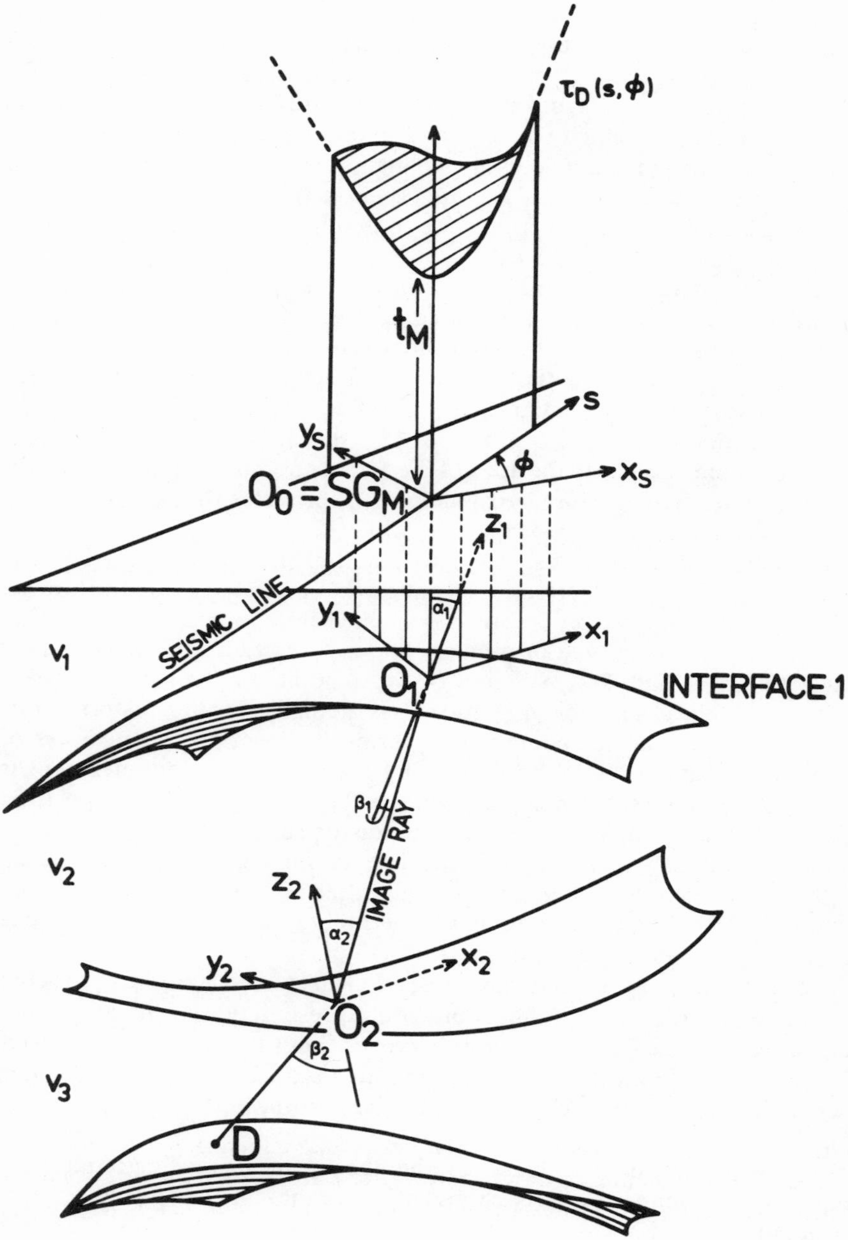

FIG. 7-4. 3-D curved layer model showing an image ray and straight seismic line through SG_M.

homogeneous wave equation may be disappointed to see their expectations occasionally not met in practice. The root of the dilemma is found in nothing else but the assumptions made about the model and the simplifications made in the theory.

Undoubtedly, various complementary approaches will be developed with the aim of providing directly from unmigrated traces truly migrated subsurface images for more or less complex velocity models. Such processes based on the wave equation have already been described (Judson et al, 1978; Schultz and Sherwood, 1978; Hatton et al, 1978). Until they, however, become standard seismic processes, time-to-depth migration (TDM) described in the next chapter, should be considered subsequent to time migration. The two TDM processes for time-migrated data that we shall describe will be referred to as *image-ray migration* and *depth migration*. Image-ray migration maps a few selected time-migrated horizons to depth whereas depth migration transforms entire time-migrated sections into depth.

Image rays play a role in forward and inverse seismic modeling of *time-migrated reflections* similar to the role played by normal incidence rays in modeling CDP-stacked reflections. Moreover, their significance for depth migrating time-migrated sections obtained by finite difference migration schemes has been demonstrated by Larner et al (1977) for the case where derivatives of the coordinate frame velocity are ignored. The amount by which image rays deviate from the true vertical is a simple and good measure of the need to perform other more sophisticated TDM processes that may be developed in the future. TDM will be discussed in chapter 8.

7.2 Small-aperture migration velocity

Since the diffraction time surface for the point scatterer D (Figure 7-3) represents twice the traveltime required by a hypothetical wavefront originating at D, we will now confine our attention to the upward traveling D-wavefront that originates at point D. Of particular interest is the traveltime behavior of this hypothetical wave in the vicinity of an image ray. This behavior is readily described by using 3-D wavefront curvature laws.

For convenience of discussion, in Figure 7-4 we show in more detail the image ray from D to the surface of the earth. 0_i designates the point where the image ray pierces the ith interface.

Consider the ray as being traced upward from D to the surface. The right-hand $[x_i, y_i, z_i]$ system may denote the $[x_F, y_F, z_F]$ system at 0_i. It is chosen such that the x_i-axis is confined to the plane of incidence for the next higher interface. The z_i-axis points upwards into the direction of the interface normal vector at 0_i. A $[x_s, y_s, z_s]$ system is placed on the image ray emergence point 0_0 (or SG_M).

Because the image ray emerges vertically, the x_s-axis and y_s-axis can be chosen at will; we choose them in such a way that they coincide with the vertical projection of the x_1- and y_1-axis at 0_1 into the earth's surface. The ith interface is approximated within the $[x_i, y_i, z_i]$ system by the following parabolic equation

$$2z_i = \mathbf{X}_i \mathbf{B}_i \mathbf{X}_i^T. \tag{7.1}$$

$\mathbf{B}_i$ is the interface curvature matrix at 0_i, and $\mathbf{X}_i = (x_i, y_i)$.

Assuming a hypothetical point source at D, we can compute the wavefront curvature matrix of the upward traveling D-wavefront with respect to the moving frame that accompanies it along the image ray. The wavefront curvature matrix of the emerging D-wavefront at 0_0 with respect to the $[x_s, y_s, z_s]$ system is $\mathbf{A}_0$. It provides the following parabolic approximation of the emerging wavefront

$$2z_s = -\mathbf{X}_s \mathbf{A}_0 \mathbf{X}_s^T. \tag{7.2}$$

The analytical expression for $\mathbf{A}_0$ is formally identical with the expression of the form (6.16), but now it pertains to the image ray rather than the normal ray. Due to the specific selection of the $[x_s, y_s, z_s]$ system, the rotation angle δ_1 is zero.

By making use of formula (4.47a) and by taking into consideration the fact that the diffraction time surface of the scatterer D at 0_N describes twice the time of the upgoing hypothetical D-wavefront, one obtains the following second-order hyperbolic approximation to the diffraction time surface for D:

$$t_H^2(x_s, y_s) = t_M^2 + \frac{2t_M}{v_1} \mathbf{X}_s \mathbf{A}_0 \mathbf{X}_s^T. \tag{7.3}$$

This is the equation of the small-aperture migration hyperboloid. The quantity t_M is the two-way time to the apex of the diffraction time curve; we have previously called it the two-way image time. The two-way image time is closely approximated by the time of the time-migrated reflection seen at 0_0; the latter strictly belongs to the apex of a migration hyperboloid that best fits the actual diffraction time surface.

Elliptical migration hyperboloids can generally be well fitted to true diffraction time surfaces within a fairly large aperture range. This conclusion is based on ray-tracing computations of diffraction time surfaces for scatterers below plane, isovelocity layers of differing, but not too steep, dip and strike. Thus, when interfaces strike in different directions, the diffraction time surfaces for (infinitely) small apertures are elliptical, rather than rotational, hyperboloids.

If a straight seismic line in the direction defined by $y_s = x_s \tan \phi$ is placed through 0_0 (Figure 7-4), then the portion of the diffraction time surface of D observed in the recorded data is a hyperbola-like curve. For small apertures it can be approximated by the following second-order equation

$$\tau_D^2(s, \phi) = t_M^2 + 4s^2/V_{\mathrm{SAM}}^2(\phi). \tag{7.4}$$

The aperture width a is twice the distance s away from 0_0; V_{SAM} is subsequently called the *small-aperture migration* (SAM) *velocity*. It depends on the azimuth ϕ. According to equation (7.3), V_{SAM} is connected with $\mathbf{A}_0$ by the following equation

$$1/V_{\mathrm{SAM}}^2(\phi) = \frac{t_M}{2v_1} \mathbf{e} \mathbf{A}_0 \mathbf{e}^T, \tag{7.5}$$

where

$$\mathbf{e} = (\cos \phi, \sin \phi).$$

V_{SAM} in time migration, is the counterpart of V_{NMO} in CDP stacking. It is associated with an image ray in the same way that V_{NMO} is associated with a normal ray. For horizontal isovelocity layers both V_{NMO} and V_{SAM} reduce to V_{RMS}. It follows that computer-derived (optimum) migration velocities (see chapter 11) depend on aperture size in much the same way that optimum stacking velocities depend on spread-length.

In CDP stacking, one must distinguish between CDP reflection time curves for primaries and those for multiples. In 3-D time migration, one likewise ought also to discriminate between diffraction time surfaces relating to primary and multiple image raypaths.

In 2-D time migration, however, one should also remain aware of diffraction curves relating to primary and multiple rays that project so as to appear like image rays in the plane of the section but actually emerge at the earth's surface outside the vertical plane of the seismic line. True migration velocities as defined above must pertain to diffraction curves whose apexes are strictly located at an emerging image ray rather than obliquely emerging scatter rays.

So long as migration velocities provide a satisfactory time migration, one can expect them to approximate V_{SAM} (the SAM velocity). There exists, however, a bias between the two quantities similar to the bias between NMO and stacking velocity. The difference between V_{SAM} and a "finite-aperture migration velocity" is called the *aperture bias*. It is equivalent to the spread-length bias in the estimation of V_{NMO} from CDP data.

From equation (7.5), one can conclude that the curvature matrix $\mathbf{A}_0$ belonging to the emerging *D*-wavefront can be recovered from SAM velocities measured in three different directions through 0_0. Processes leading to optimum time-migrated reflection data can thus provide $\mathbf{A}_0$ along with the two-way image time to all subsurface scatterers.

Knowing both these quantities for a scatterer D on the Nth reflector and interval velocities computed for the upper $N - 1$ layers, one can compute v_N in a noniterative manner and locate D in 3-D space. The procedure is developed in chapter 9. Again the algorithm will be based on the concept of a *D*-wavefront shrinking back into its hypothetical source at D.

In a 2-D situation where image and normal rays fall into a vertical seismic plane, it is sufficient to know V_{SAM} and t_M for each time-migrated reflection element in this plane to recover interval velocities, depths, and shapes of reflectors.

7.2.1 Summary and perspectives

The traveltimes of a CDP-stacked primary reflecting time horizon (e.g., as given in the form of a contoured time map) are closely approximated by the two-way normal reflection times for coincident source-receiver pairs at CDP positions. The traveltimes are computed along normal rays.

The two-way time along a scatter ray from a coincident source-receiver pair to a selected subsurface scattering point is confined to the diffraction time surface of that point. This surface generally has a hyperboloid-like shape in the vicinity of the apex. Each subsurface point is normally associated with

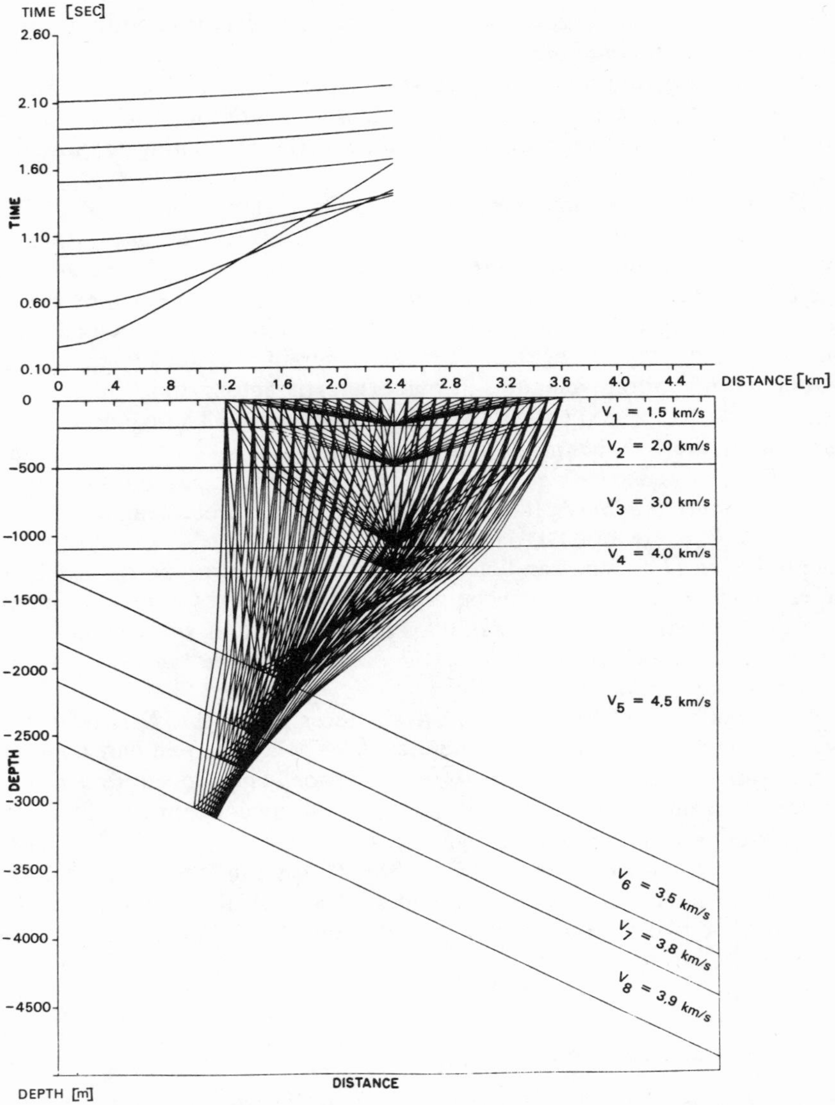

FIG. 7-5. 2-D plane-dipping layer model with CDP ray families and CDP reflection time curves to all reflecting velocity boundaries.

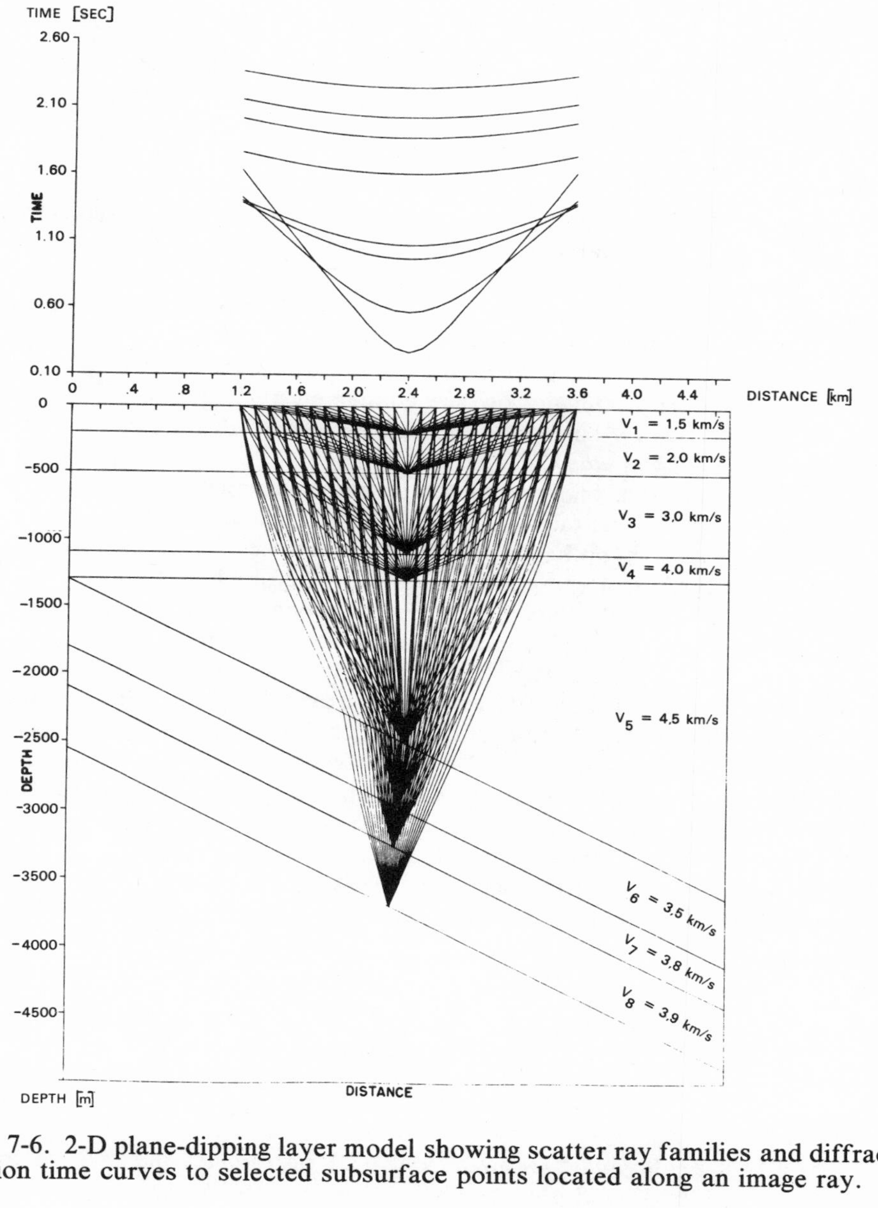

FIG. 7-6. 2-D plane-dipping layer model showing scatter ray families and diffraction time curves to selected subsurface points located along an image ray.

Table 7-1. Model parameters.

	Horizon number							
	1	2	3	4	5	6	7	8
Depth (m)	200	500	1100	1300	1305	1800	2100	2550
Vel (m/sec)	1500	2000	3000	4000	4500	3500	3800	3900
Dip (m/1000 m)	0	0	0	0	500	500	500	500

a different diffraction time surface; typically that surface will have only one apex coinciding with the minimum of the time surface.

Time migration by definition sums scattered signals into the minimum time point. A *two-way normal reflection time map* is thereby transformed into a *two-way image time map*. In effect, considering all diffraction time surfaces that are tangent to a reflecting horizon on a stacked CDP section, the process repositions the horizon along a surface connecting the apexes (or minima) of these surfaces. The new surface is the *time-migrated horizon* or, equivalently, the *two-way image time map*.

There is an interesting aside to this particular transformation of time horizons (from unmigrated to time-migrated maps). It is obvious that image rays can be associated with a hypothetical plane wave that begins at time zero at the surface of the earth and propagates downward before being scattered. Since image rays already describe the downward wave propagation paths, a time section resulting from an actual (or simulated) plane surface source can, in a sense, be considered as being already "half-migrated." As each reflecting horizon constitutes a continuum of scatterers, the apexes of resulting diffraction time surfaces in a "plane wave time section" will coincide with image-ray locations at the surface. Time migration performed on such a section will consequently place primary time horizons at the same positions as those picked from a time-migrated conventional CDP-stacked section. Consequently, wherever velocities vary laterally, TDM might be necessary here also.

If the local velocity distribution above a particular reflector is constant or only vertically inhomogeneous, image rays are vertical in their whole length,

Table 7-2. Results of the model for Table 7-1.

	Horizon number							
	1	2	3	4	5	6	7	8
Normal time	0.267	0.567	0.967	1.067	1.510	1.760	1.904	2.111
Image time	0.267	0.567	0.967	1.067	1.602	1.872	2.025	2.249
V_{NMO}	1500	1782	2363	2562	3490	3554	3610	3638
V_S	1500	1843	2444	2655	3531	3577	3612	3645
V_{SAM}	1500	1782	2363	2562	3337	3337	3363	3407
V_M	1500	1843	2443	2655	3358	3347	3368	3409

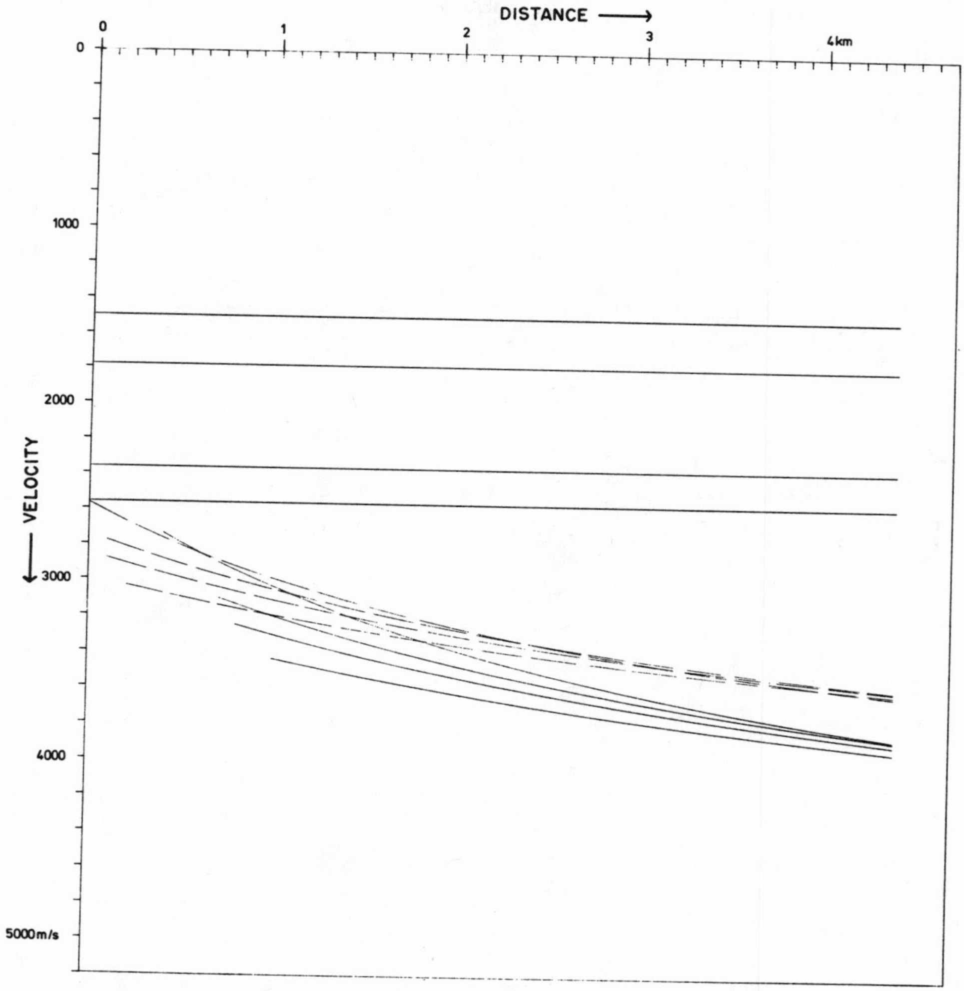

FIG. 7-7. Plot of V_{NMO} (solid lines) and V_{SAM} (dashed lines) as a function of CDP points along a seismic line.

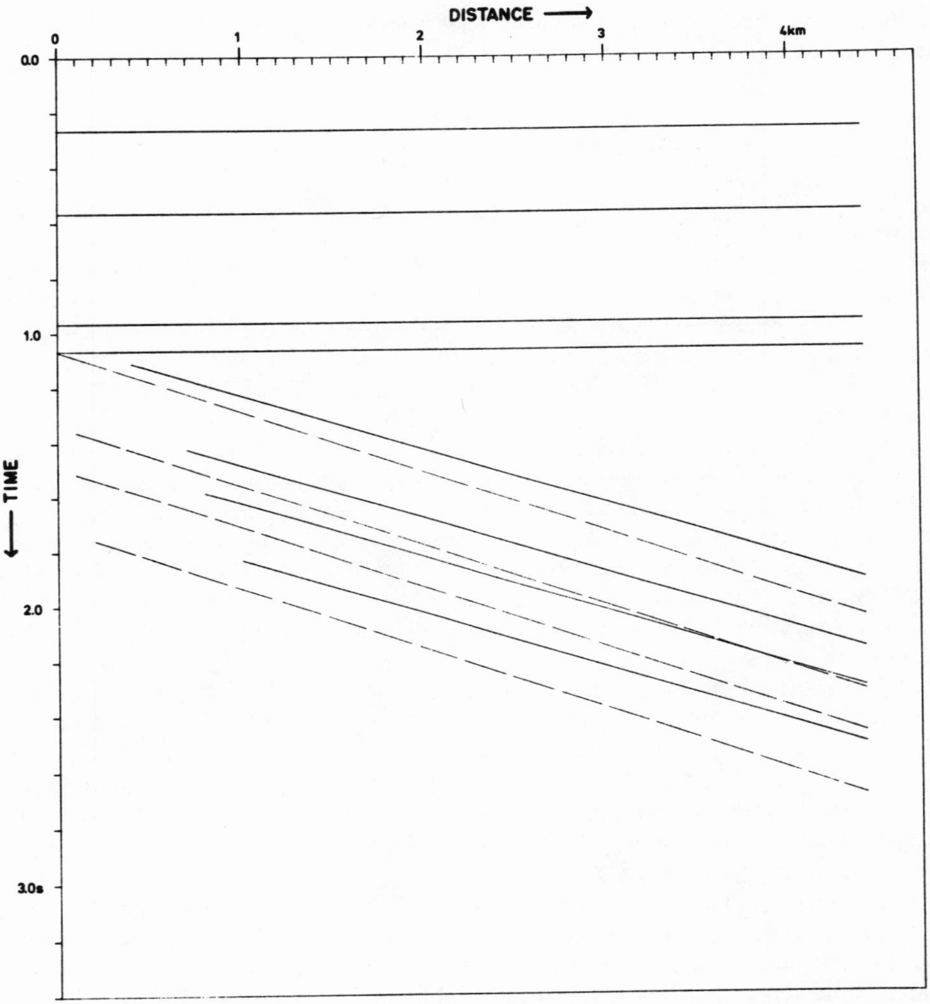

FIG. 7-8. Two-way normal (solid lines) and two-way image times (dashed lines) to reflecting plane velocity boundaries.

and time migration provides a complete migration. In areas of complex geology where velocities vary laterally, time migration might, however, have to be followed by an additional TDM if the true shape of the depth reflectors is to be reconstructed.

Diffraction time surfaces can be approximated in much the same way as are CDP reflection time curves. The image ray assumes a role comparable to that of the normal ray; V_{SAM} is the counterpart of V_{NMO}; migration velocity is the counterpart of stacking velocity; and aperture half-width is the counterpart of maximum offset.

7.3 Modeling of velocities

Much can be learned from modeling normal and image rays (or normal and image times) for a given subsurface velocity model. Likewise, modeling the parameters V_{NMO} and V_{SAM} is useful. Figure 7-5 shows a 2-D plane-dipping, constant-layer velocity model covered by a 24-fold, CDP spread centered at $X = 2400$ m. The model parameters are given in Table 7-1.

CDP ray families are traced to all interfaces. The upper portion of Figure 7-5 shows computed NMO curves for this model. Table 7-2 lists computed normal times, as well as various velocities at position $X = 2400$ m. V_S was computed by least-square fitting a hyperbola (with a fixed apex position at offset zero) to the NMO curves.

Figure 7-6 shows the same velocity model with an image ray emerging at $X = 2400$ m. Also shown are the scatter rays from points of intersection between the image ray and the velocity boundaries. The diffraction time curves all have their minima at the same emerging image-ray location. The values for V_{SAM} and the migration velocity V_M were established for these interface points. The results are given in Table 7-2.

V_M is obtained by least-square fitting an exact hyperbola (with a fixed apex position at $X = 2400$ m) to the diffraction time curve. Note the differences in parameters for the dipping layers 5–8.

If V_S and V_M need to be accurate to within ±5 percent, then clearly V_S and V_M do not adequately approximate each other; in laterally inhomogeneous media, these can be wholly different quantities. In contrast, V_{NMO} and V_{SAM} are useful approximations for V_S and V_M, respectively.

It is often helpful and economical to compute both V_{NMO} and V_{SAM} continuously as a function of CDP locations for selected reflectors. This is done in Figure 7-7 for the model shown in Figure 7-5. The corresponding two-way normal times (solid lines) and image times (dashed lines) are shown in Figure 7-8.

V_{NMO} and V_{SAM} are identical for the first four interfaces. For the dipping reflectors, however, they are considerably different from one another. Note that V_{NMO} and V_{SAM} do not increase linearly in direction of dip though the reflecting interfaces do. The radii of curvature for the emerging NIP wavefront or D-wavefront do vary linearly, however. (Recall also that the radius of curvature is proportional to $V^2_{NMO}\, t_N$ or $V^2_{SAM}\, t_M$ as the case may be. t_N is the normal and t_M the image time.)

An easy rule to remember is that, for a given CDP point, the two-way normal time to a reflector is always smaller than the two-way image time. This rule

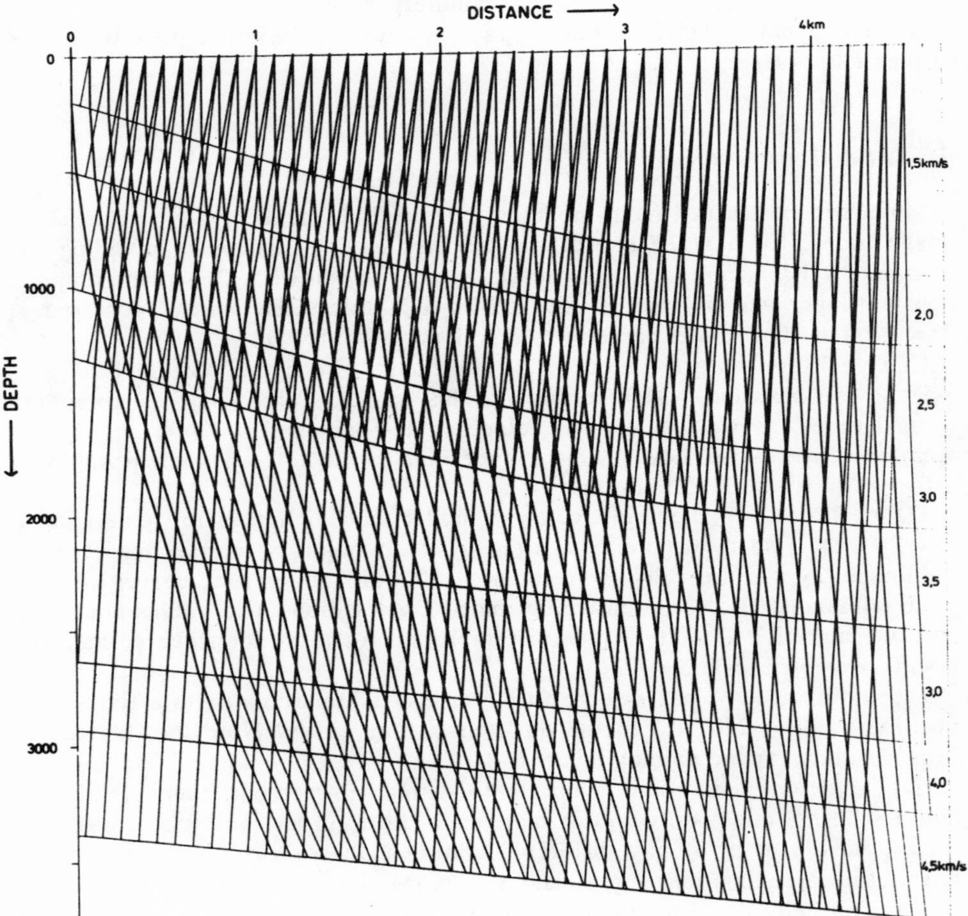

FIG. 7-9. 2-D curved layer model featuring a situation where image rays to reflecting layer boundaries deviate more from the vertical than normal rays.

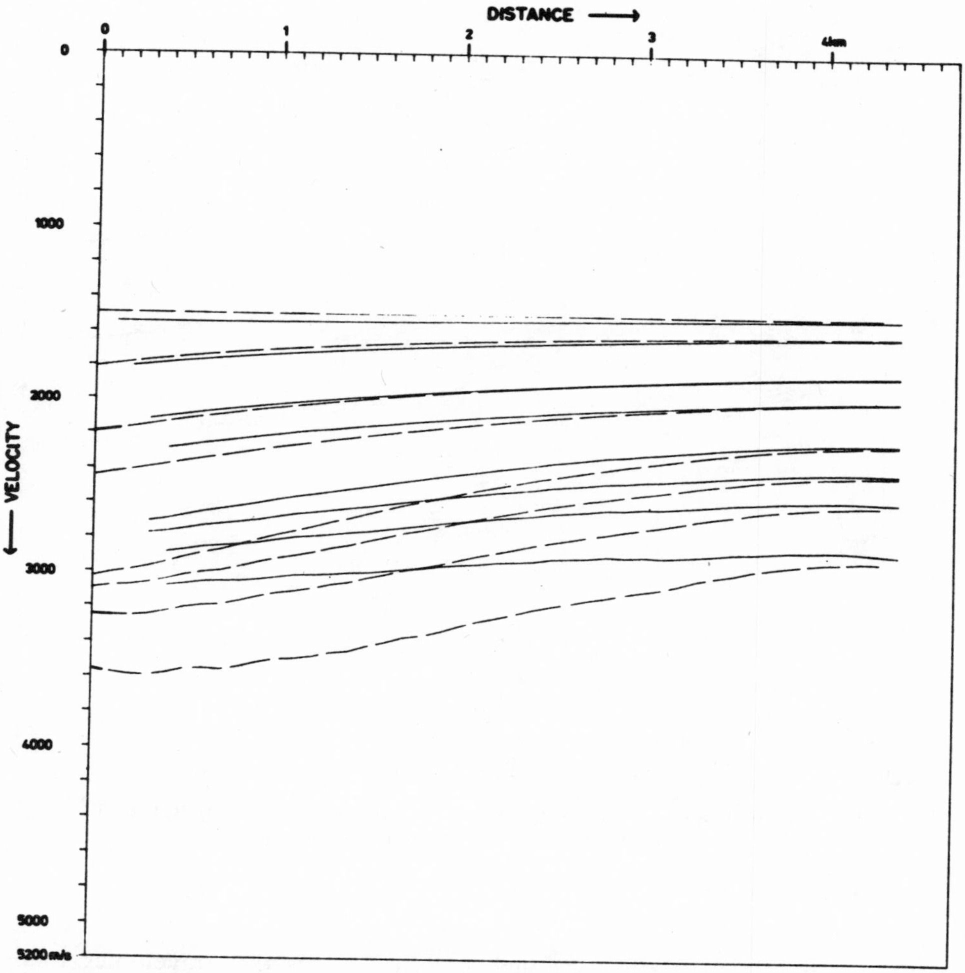

FIG. 7-10. Plot of V_{NMO} (solid lines) and V_{SAM} (dashed lines) for layer boundaries of Figure 7-9.

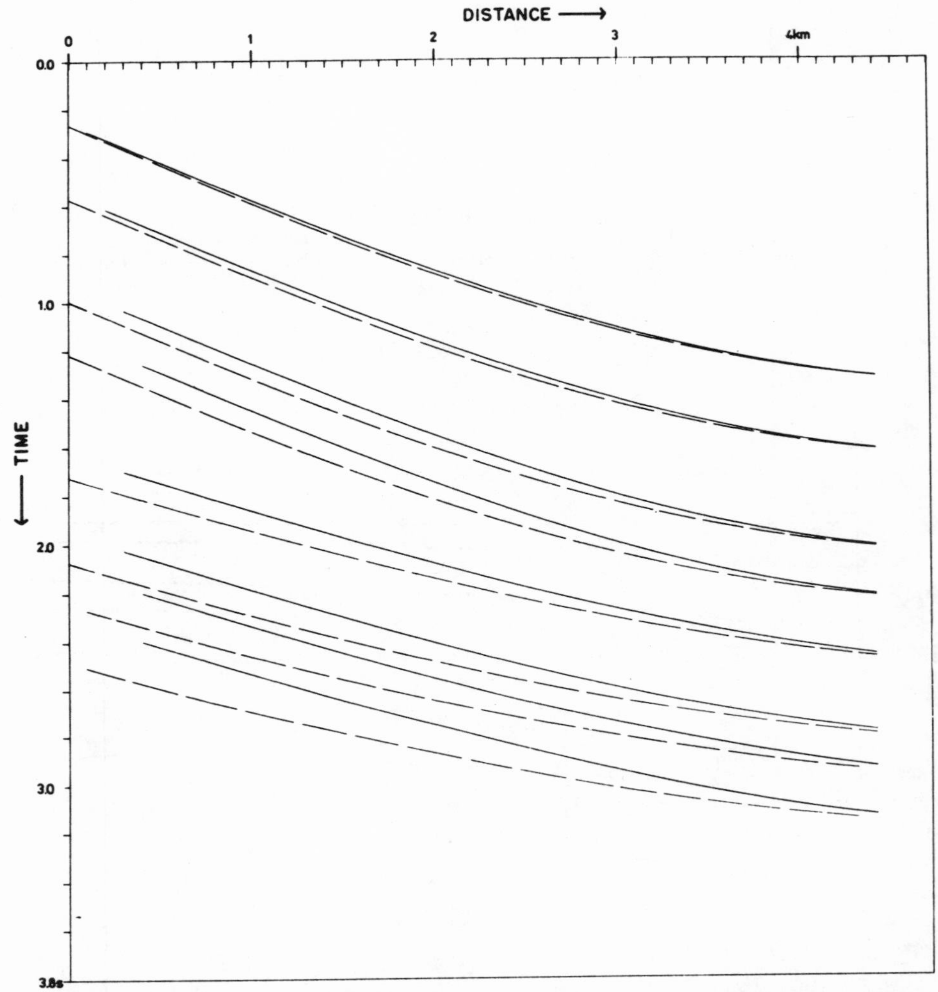

FIG. 7-11. Plot of two-way normal (solid lines) and two-way image times (dashed lines) to reflecting layer boundaries.

follows simply from Fermat's principle, which states that traveltimes follow a minimum (stationary) traveltime path.

A model similar to the one above but with curved velocity boundaries is shown in Figure 7-9. This figure features normal and image rays to the layer boundaries from CDP points that are separated by $\Delta X = 100$ m. Reflections on the time-migrated section for this earth model will actually be more poorly positioned than those on the CDP-stacked section. That is, normal rays are more nearly vertical than image rays.

Figure 7-10 displays V_{NMO} (solid lines) and V_{SAM} (dashed lines) for the velocity boundaries. Figure 7-11 shows the associated two-way normal times (solid lines) and two-way image times (dashed lines).

8 Time-to-depth migration (TDM)

Traveltimes of CDP-stacked or time-migrated reflections tempt seismic interpreters to associate them directly with shapes of the reflecting horizons in depth. Such an association is certainly justified in areas of moderate geologic complexity. Moreover, time-migrated reflections often provide a good approximation of the true shape of seismic reflectors.

In the last two chapters, however, we showed that time-migrated seismic reflections, like CDP-stacked reflections, rarely fall precisely above their corresponding depth points. Lateral position errors in time migration are caused by lateral velocity changes above the reflectors of interest. We will refer, in general, to any of the many processes that transform from apparent positions on seismic time sections (whether stacked or time-migrated) to true positions in depth as *time-to-depth migration* (TDM).

It has long been recognized (Slotnick, 1936, 1959; Rice, 1949; Krey, 1951; Hagedoorn, 1954; Musgrave, 1964) that TDM is required for unmigrated data such as seismic records and sections of CDP-stacked or normal reflections. That TDM is also required for time-migrated reflections some readers may have realized (perhaps with some disappointment) only upon reading chapter 7. Whenever normal-incidence rays or image rays deviate from the vertical, a time-to-depth conversion that follows a CDP stack or time migration must have some form of TDM as a final step. In section 8.1, we briefly describe a TDM process for normal reflections. In sections 8.2 and 8.3, we show how to perform TDM for time-migrated reflections.

As we shall see in Chapter 9, the subject of computing interval velocities is also very closely related to TDM.

In general, since complex geology is rarely strictly 2-D, TDM should be performed in three dimensions. The commonly applied conventional migration of CDP-stacked time sections (two-dimensional migration) is valid only for (virtually nonexistent) cases where all the seismic interfaces are perpendicular to one and the same plane—usually the vertical plane—through the seismic line.

To perform TDM strictly requires knowledge of the velocity distribution in the subsurface. Depending upon the form in which this velocity distribution is available and upon the complexity of the subsurface model, TDM of CDP stacked reflections can involve some more or less complicated ray-tracing or map transformation procedures (Sattlegger, 1964, 1969, 1977; Kleyn, 1977; Michaels, 1977).

We confine further considerations again to the familiar 3-D isovelocity layer model of Figure 2-1 and describe a 3-D TDM scheme that involves the recursive

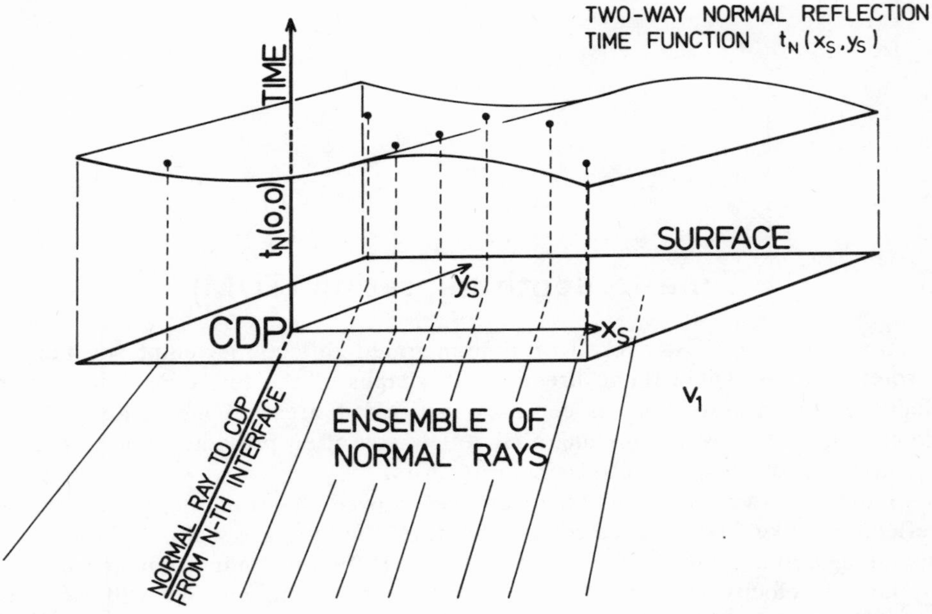

FIG. 8-1. Plot showing ensemble of emerging normal rays and the two-way normal time function for selected reflector.

ray tracing and curvature algorithms described earlier. The model is assumed to approximate the velocity distribution of the geology in the vicinity of each CDP. The scheme is used later mainly for recovering (reflecting) first-order velocity boundaries when computing interval velocities.

In the development below, the boundaries of the upper $N-1$ velocity layers are considered to have already been computed. The observed Nth reflection time map is available in either a CDP-stacked or time-migrated form. TDM mapping procedures convert the Nth time map into the Nth horizon in depth. Though not necessary for the solution of the problem, curvatures of normal wavefronts will be included in two of the TDM schemes described in the following because they provide additionally the curvature of the reflecting beds, required later for computing interval velocities of still deeper horizons.

8.1 Mapping CDP-stacked reflections

The mapping of CDP-stacked reflections into depth is achieved with the help of normal rays. The process is also referred to as *normal ray migration.* Figure 8-1 features a normal ray from the CDP to the NIP, one of an ensemble of normal rays traced from the surface to the Nth reflector. As indicated above, normal rays can be associated with a hypothetical normal wavefront that originates at the selected reflecting interface and propagates upward to the surface of the earth. The traveltimes belonging to all normal rays in the vicinity of the ray from the NIP to the CDP can be approximated with the help of wavefront curvature laws.

For a wavefront propagating upward, the z-axis of the moving $[x, y, z]$ system and the z_F-axes of the $[x_F, y_F, z_F]$ systems at interfaces point upward. Let $\mathbf{B}_N$ be the interface curvature matrix of the Nth reflector at the NIP. It equals the negative wavefront curvature matrix of the hypothetical normal wavefront that originates there. In the following we ignore specification of the quantities in the recursion (8.1). However, the wavefront curvature matrix A_0 of the emerging normal wavefront at CDP can be obtained by performing a recursion in proper analogy to recursion (4.40) along the normal ray from the NIP to the CDP:

$$\begin{aligned}
\mathbf{A}_{I,N-1}^{-1} &= -\mathbf{B}_N^{-1} + s_N \mathbf{I} \\
\mathbf{A}_{T,N-1} &= \mathbf{D}_{N-1}^{-1} \left[\frac{v_{N-1}}{v_N} \mathbf{S}_{N-1} \mathbf{A}_{I,N-1} \mathbf{S}_{N-1} + \rho_{N-1} \mathbf{S}_{T,N-1}^{-1} \mathbf{B}_{N-1} \mathbf{S}_{T,N-1}^{-1} \right] \mathbf{D}_{N-1} \\
\mathbf{A}_{I,N-2}^{-1} &= \mathbf{A}_{T,N-1}^{-1} + s_{N-1} \mathbf{I} \\
&\cdots\cdots\cdots\cdots\cdots\cdots \\
&\cdots\cdots\cdots\cdots\cdots\cdots \\
\mathbf{A}_0^{-1} &= \mathbf{A}_{T,1}^{-1} + s_1 \mathbf{I}.
\end{aligned} \tag{8.1}$$

For our purposes here, we are assuming that all details of the layering are known down to, but not including, the Nth interface. In particular, the interface curvature matrices $\mathbf{B}_j$ ($j = 1, 2, \ldots, N - 1$) are available. They must have been obtained by interpolation of curvatures observed at neighboring points along the respective interfaces (because, in general, NIPs for different interfaces lie along different rays) or by fitting a surface to the computed NIPs for different CDP locations. Stated differently, a characteristic of the NIP approach is that a surface must be fitted to the points defined for interface $j - 1$ before proceeding toward a solution for interface j.

According to equation (4.47), the two-way normal reflection time map expressed in the $[x_s, y_s]$ system at the CDP can then be approximated by the following parabolic equation:

$$t_N(x_s, y_s) = t_N(0, 0) + \frac{2 \sin \beta_0}{v_1} x_s + \frac{1}{v_1} \mathbf{X}_s \mathbf{S}_0 \mathbf{A}_0 \mathbf{S}_0 \mathbf{X}_s^T. \tag{8.2}$$

$t_N(0, 0)$ is the two-way normal time from the CDP to the Nth horizon.

Conventional seismic profiling provides two-way normal times only along one direction—that of the seismic line. Other quasi-linear field methods such as zig-zag lines or shooting with various lateral offsets (e.g., wide line profiling, crooked lines, slalom lines) provide these times within a strip on both sides of a fictitious seismic line. If conventional seismic lines are sufficiently close to one another, two-way normal reflection time functions are available from contoured normal time maps.

From the linear term of formula (8.2) (compare formula 4.30) one obtains the direction of the emerging ray. By using the ray-tracing technique described in section 4.2, one can then trace the normal ray to $\mathbf{O}_N$ from CDP down to $\mathbf{O}_{N-1}$.

Let t_{N-1} be the two-way time from CDP to $\mathbf{O}_{N-1}$ via $\mathbf{O}_1$, $\mathbf{O}_2$, etc. Then $\Delta t_N = t_N(0,0) - t_{N-1}$ is the final two-way time in the Nth velocity layer, and $s_N = \Delta t_N \, v_N/2$ is the distance from $\mathbf{O}_{N-1}$ to NIP. If $\mathbf{e}$ is the directional unit vector from $\mathbf{O}_{N-1}$ to $\mathbf{O}_N$, one can obtain the position of NIP from

$$\mathbf{O}_N = \mathbf{O}_{N-1} + s_N \, \mathbf{e}, \tag{8.3}$$

where $\mathbf{e}$ is determined from Snell's law. It is available at the $(N-1)$th velocity boundary as v_{N-1} and v_N are assumed to be known.

In addition to defining the Nth interface only by the lower end points of normal rays, one may wish to determine $\mathbf{B}_N$ at these points. As $\mathbf{A}_0$ in equation (8.2) is available from surface measurements at the CDP, recursion (8.1) need only be reversed to provide the interface curvature matrix $\mathbf{B}_N$ and thus a second-order approximation to the reflecting interface at NIP.

This process can be viewed physically in terms of having the normal wavefront shrink back into the reflecting interface. TDM in this sense can be viewed as a downward continuation process—a reversal of a hypothetical wave propagation. Mathematically, this reversal can be achieved in either of two different ways. In one, the normal wave propagates *backward* in time using the coordinate frame of the hypothetical upward moving normal wavefront. This is formally achieved by a simple reversal of the recursion (8.1).

The other way (which we recommend here) is to view the downgoing wavefront as a *forward* propagating with respect to a *positively growing* time. In this latter case, one may use the standard ray-tracing algorithm of section 4.2, start at time zero, use $-\mathbf{A}_0$ instead of $\mathbf{A}_0$ [which would result from formula (8.2)], and describe the downgoing normal wavefront within a moving $[x, y, z]$ frame where the z-axis points into the direction of the *downward* traced normal ray.

In summary, the position of $\mathbf{O}_N$ for a given velocity field is defined entirely by the normal time and its lateral gradient at CDP. But additionally, the curvature matrix $\mathbf{B}_N$ at NIP, useful for contouring, can be computed if second-order derivatives of the Nth normal time map can be established at CDP.

For smooth depth horizons, the quantities $\mathbf{B}_N$ and the reflection points of normal rays for a set of CDP points should vary smoothly in all lateral directions. While working downward from top to bottom, layer by layer, one can, uncertain observations of traveltime and velocity, use this constraint to check both. TDM can thus provide a tool for testing the quality of traveltimes and interval velocities. interval velocities.

8.2 Mapping time-migrated reflections

The mapping of time-migrated reflections into depth is achieved with the help of image rays. The process is also referred to as *image-ray migration.* Figure 8-2 shows an image ray at surface point SG_M that is assumed to belong to a scatterer D on the Nth reflector. If v_i $(i = 1, \ldots, N)$ and the upper N-1 velocity boundaries are known, then the position of D can be easily found from knowledge of only the two-way image time of the time-migrated reflection observed at SG_M.

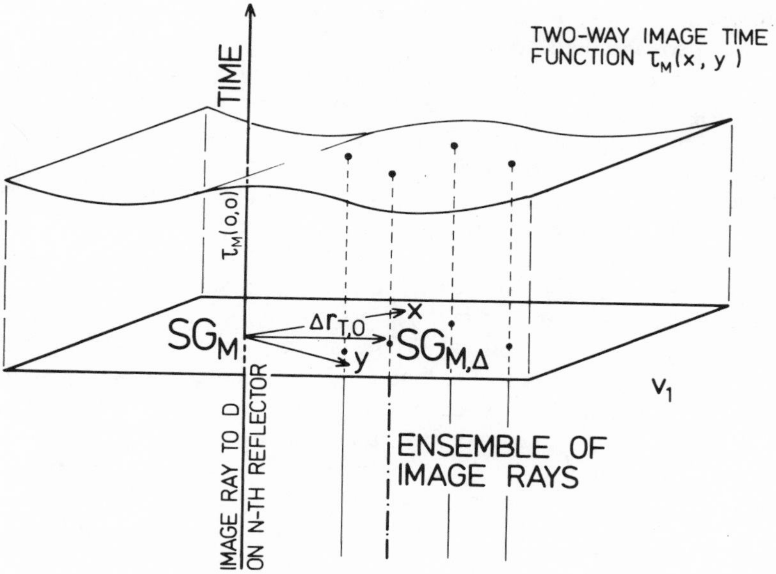

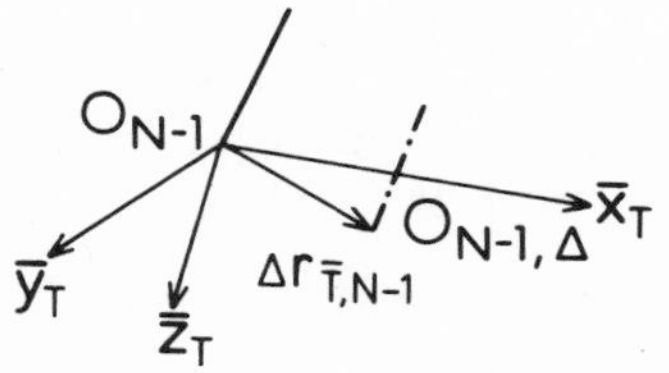

FIG. 8-2. Plot showing ensemble of emerging image rays and the two-way image time function for a selected interface. Sketched below is the $[\bar{x}_T, \bar{y}_T, \bar{z}_T]$ coordinate system at subsurface point 0_{N-1} with a portion of the image ray that passes through $SG_{M,\Delta}$.

All one needs to do is to trace the image ray downward by starting vertically from the surface of the earth. The procedure is similar to that described above for finding the position of the NIP by downward tracing a normal ray.

The position of D depends only on the two-way image time. It is found by tracing the image ray down from SG_M until half of the observed two-way image time is consumed. An advantage in working with image rays rather than normal rays is that uncertainties in observed time gradients for steeply dipping reflectors will generally influence the result less. That is, depth-migrated positions of steeply dipping reflectors are less sensitive to uncertainties in the time gradients in time-migrated data than in those of unmigrated (i.e., CDP-stacked or normal-time) data.

The TDM method presented in section 8.1 employed the concept of a hypothetical normal wavefront that originates within the Nth interface. The method was concerned largely with computing the interface curvature matrix

$\mathbf{B}_N$. In a similar way as $\mathbf{B}_N$ can be derived from two-way normal time maps, it can be also obtained from two-way image time maps. The theory required to establish this second approach is more complex, but the method turns out to be computationally much more efficient. It has the advantage that solutions can be obtained at individual SG_M points, independent of others. That is, all interface curvature matrices $\mathbf{B}_j$ ($j = 1, \ldots, N$) are computed at intersection points along the same image ray; no fitting of a surface to interface $j - 1$ is required before proceeding to a solution for interface j. The method makes use of second-order derivatives of the two-way image time map of the Nth reflecting horizon and involves a hypothetical wavefront that can be associated with image rays. The method is described in the following.

Since image rays are vertical at the surface, one can associate them with a hypothetical plane wavefront that originates along the surface of the earth at a given time instant. We will call this hypothetical downward moving wavefront the *image wavefront*. The two-way times required for an image wavefront to travel between points on the earth's surface and on the Nth reflector define the two-way image time map, so long as times are plotted at emergence points of image rays on the earth's surface. The image wavefront can be accompanied by the familiar moving $[x, y, z]$ frame along a selected image ray between SG_M and D.

From within the ensemble of image rays, let us choose an image ray close to the one from SG_M to D, i.e., the ray that connects the points $SG_{M,\Delta}$ and D_Δ. Both D and D_Δ are on the reflecting (or better, scattering) depth horizon.

Within the $[\bar{x}_T, \bar{y}_T, \bar{z}_T]$ system on the lower side of point $\mathbf{O}_{N-1}$ (see Figure 8.2), one can approximate the Nth as yet unknown reflecting interface as

$$\bar{z}_T(\bar{x}_T, \bar{y}_T) = s_N + \bar{\mathbf{X}}_T \bar{\mathbf{C}}^T + \bar{\mathbf{X}}_T \bar{\mathbf{B}}\, \bar{\mathbf{X}}_T^T; \tag{8.4}$$

$$\bar{\mathbf{X}}_T = (\bar{x}_T, \bar{y}_T) \quad ; \quad \bar{\mathbf{C}} = (\bar{c}_1, \bar{c}_2);$$

$$\bar{\mathbf{B}} = \begin{bmatrix} \bar{b}_{11} & \bar{b}_{12} \\ \bar{b}_{12} & \bar{b}_{22} \end{bmatrix}.$$

As will be shown in the following, the quantities s_N, $\bar{\mathbf{C}}$, and $\bar{\mathbf{B}}$ are contained in the first- and second-order derivatives of the Nth two-way image time map at SG_M, assumed to be expandable in the following form

$$\tau_M(x, y) = \tau_M(\mathbf{X}) = t_N + \mathbf{X}\mathbf{E}^T + \mathbf{X}\mathbf{F}\mathbf{X}^T, \tag{8.5}$$

where

$$\mathbf{X} = (x, y) \quad ; \quad \mathbf{E} = (e_1, e_2) \quad ; \quad \mathbf{F} = \begin{bmatrix} f_{11} & f_{12} \\ f_{12} & f_{22} \end{bmatrix},$$

and t_N is the two-way image time from SG_M to D; the quantities $\mathbf{E}$ and $\mathbf{F}$ are assumed to have been estimated from the image time map at SG_M (e.g., by some least-square or spline-fitting method).

Note that the surface coordinate system $[x, y]$ used here is part of the right-hand $[x, y, z]$ frame that moves downward along the image ray from SG_M to D. The x-axis is chosen in such a way that it lies in the plane of incidence at $\mathbf{O}_1$.

We now compute the wavefront curvature matrix $\mathbf{A}_{\bar{T},N-1}$ of the hypothetical downgoing image wavefront at the lower side of $\mathbf{O}_{N-1}$ within the $\bar{x}_T - \bar{y}_T$ plane of the local $[\bar{x}_T, \bar{y}_T, \bar{z}_T]$ system. The matrix is obtained by way of the following recursion.

1) Transmitted side of $\mathbf{O}_0$ $(=SG_M)$

$$\mathbf{A}_{T,0} = \mathbf{N} = \begin{bmatrix} 0 & 0 \\ 0 & 0 \end{bmatrix}.$$

2) Incident side of $\mathbf{O}_1$

$$\mathbf{A}_{I,1} = \mathbf{N}.$$

3) Transmitted side of $\mathbf{O}_1$ due to equation (4.38)

$$\mathbf{A}_{T,1} = \mathbf{D}_1^{-1} \, [(v_2/v_1)\, \mathbf{S}_1 \mathbf{A}_{I,1} \mathbf{S}_1 + \rho_1 \, \mathbf{S}_{T,1}^{-1} \, \mathbf{B}_1 \, \mathbf{S}_{T,1}^{-1}] \, \mathbf{D}_1.$$

4) Incident side of $\mathbf{O}_2$

$$\mathbf{A}_{I,2}^{-1} = \mathbf{A}_{T,1}^{-1} + s_2 \mathbf{I}.$$

Continuing the recursion, we find $\mathbf{A}_{\bar{T},N-1}$ in the form

$$\mathbf{A}_{\bar{T},N-1} = [(v_N/v_{N-1})\, \mathbf{S}_{N-1} \mathbf{A}_{I,N-1} \mathbf{S}_{N-1} + \rho_{N-1} \, \mathbf{S}_{T,N-1}^{-1} \, \mathbf{B}_{N-1} \, \mathbf{S}_{T,N-1}^{-1}].$$

Let us now assume that the $\bar{x}_T - \bar{y}_T$ plane at $\mathbf{O}_{N-1}$ is densely covered with receivers which record the arriving image wavefront. A second-order parabolic approximation to the traveltime within this plane is then

$$\tau_{N-1}(\bar{x}_T, \bar{y}_T) = t_{N-1} + \frac{1}{2 v_N} \bar{\mathbf{X}}_T \mathbf{A}_{\bar{T},N-1} \bar{\mathbf{X}}_T^T, \tag{8.6}$$

where t_{N-1} is the one-way image time from SG_M to $\mathbf{O}_{N-1}$.

To equation (8.6) one has to add the extra time needed by the image wavefront to reach the reflecting Nth interface. Using equation (8.4), one obtains for the additional time required by the image wavefront to reach the Nth interface:

$$\tau_c(\bar{x}_T, \bar{y}_T) = \frac{1}{v_N} (s_N + \bar{\mathbf{X}}_T \bar{\mathbf{C}}^T + \bar{\mathbf{X}}_T \bar{\mathbf{B}} \bar{\mathbf{X}}_T^T). \tag{8.7}$$

The total two-way time τ_I along image rays can thus be approximated within the $\bar{x}_T - \bar{y}_T$ plane at $\mathbf{O}_{N-1}$ as

$$\tau_I(\bar{x}_T, \bar{y}_T) = 2\, [\tau_{N-1}(\bar{x}_T, \bar{y}_T) + \tau_c(\bar{x}_T, \bar{y}_T)]. \tag{8.8}$$

We finally must relate $\tau_I(\bar{x}_T, \bar{y}_T)$ of equation (8.8) to $\tau_M(x, y)$ of equation (8.5) in order to obtain a second-order approximation of the interface at D.

Let $\Delta\mathbf{r}_{T,0}$ be the infinitesimal vector pointing from SG_M to the nearby point $SG_{M,\Delta}$ on the surface of the earth. The point where the image ray from $SG_{M,\Delta}$ pierces the $\bar{x}_T - \bar{y}_T$ plane at $\mathbf{O}_{N-1,\Delta}$ is described by the vector $\Delta\mathbf{r}_{\bar{T},N-1}$ with respect to the $[\bar{x}_T, \bar{y}_T, \bar{z}_T]$ system. It is obvious that the following condition must hold true

$$\tau_M(\Delta\mathbf{r}_{T,0}) = \tau_I(\Delta\mathbf{r}_{\bar{T},N-1}). \qquad (8.9)$$

$\Delta\mathbf{r}_{\bar{T},N-1}$ can be found from $\Delta\mathbf{r}_{T,0}$ by a recursion similar to that used to obtain $\mathbf{A}_{\bar{T},N-1}$ from $\mathbf{A}_{T,0}$. This recursion is described next.

Using the same logic as applied in Appendix C and neglecting terms of second and higher order in $|\Delta\mathbf{r}_{T,0}|$, one obtains for the first three recursive steps while proceeding from the surface downward:

(1) $\Delta\mathbf{r}_{I,1} \approx \Delta\mathbf{r}_{T,0} + s_1\,\Delta\mathbf{r}_{T,0}\,\mathbf{A}_{T,0} = \Delta\mathbf{r}_{T,0}(\mathbf{I} + s_1\,\mathbf{A}_{T,0})$,

(2) $\Delta\mathbf{r}_{T,1} \approx \Delta\mathbf{r}_{I,1}\,\mathbf{S}_1^{-1}\,\mathbf{D}_1^{-1} = \Delta\mathbf{r}_{T,0}(\mathbf{I} + s_1\,\mathbf{A}_{T,0})\,\mathbf{S}_1^{-1}\,\mathbf{D}_1^{-1}$,

and

(3) $\Delta\mathbf{r}_{I,2} \approx \Delta\mathbf{r}_{T,1}(\mathbf{I} + s_2\,\mathbf{A}_{T,1}) = \Delta\mathbf{r}_{T,0}(\mathbf{I} + s_1\,\mathbf{A}_{T,0})\,\mathbf{S}_1^{-1}\,\mathbf{D}_1^{-1}(\mathbf{I} + s_2\,\mathbf{A}_{T,1})$.

By mathematical induction, one arrives at

$$\Delta\mathbf{r}_{\bar{T},N-1} \approx \Delta\mathbf{r}_{T,0}\,\mathbf{G}, \qquad (8.10)$$

where

$$\mathbf{G} = \prod_{j=1}^{N-1}(\mathbf{I} + s_j\,\mathbf{A}_{T,j-1})\,\mathbf{S}_j^{-1}\,\mathbf{D}_j^{-1}, \qquad (8.11)$$

and

$$\mathbf{D}_{N-1} = \mathbf{I}.$$

$\mathbf{G}$ and $\mathbf{A}_{T,j}$ ($j = 1, \ldots, N-1$) are defined entirely by seismic parameters along the image ray from SG_M to $\mathbf{O}_{N-1}$. For the definitions of $\mathbf{S}_j$ and $\mathbf{D}_j$, see section 4.3.5.

Substituting equations (8.10) and (8.5) into equation (8.9) and accounting for equations (8.6) to (8.8) provides, in the limit as $\Delta\mathbf{r}_{T,0}$ approaches zero, expressions for the desired quantities $\bar{\mathbf{C}}$ and $\bar{\mathbf{B}}$ of equation (8.4):

$$\bar{\mathbf{C}}^T = \frac{1}{2}\mathbf{G}^{-1}\,\mathbf{E}^T, \qquad (8.12)$$

and

$$\bar{\mathbf{B}} = \frac{1}{2}\left(\mathbf{G}^{-1}\,\mathbf{F}\,(\mathbf{G}^{-1})^T - \frac{1}{v_N}\mathbf{A}_{\bar{T},N-1}\right). \qquad (8.13)$$

Finally, with the help of equation (A.11), one can obtain $\mathbf{B}_N$ from $\bar{\mathbf{B}}$.

Image rays to the $(N+1)$th reflector include image rays to the Nth reflector. Thus using the above described procedure, we can immediately use $\mathbf{B}_N$ to find $\mathbf{B}_{N+1}$ from the two-way image time function for the $(N+1)$th reflector at SG_M, etc.

In addition to the two TDM mapping schemes described above, let us now describe a TDM scheme by which all time-migrated data are transformed to depth. The process is referred to as *depth migration*. Unlike the above two methods, it will not enter the schemes of computing interval velocities described later. It is, however, a very useful method in connection with displaying seismic data.

8.3 Depth migration

The process of time migration is a powerful imaging tool that generally leads to more useful images of the earth than are given by CDP-stacked sections. Because imaging principles, inherent in obtaining a time-migrated section, are generally incomplete when velocities vary laterally, one may consider a follow-up process that migrates *all* time-migrated reflections into truly migrated depth locations.

Whereas image-ray migration, as discussed above, aims essentially at finding the true depth locations of contoured image time maps for selected key horizons, seismic depth migration involves the transfer of all samples of a time-migrated seismic section into depth.

Depth migration requires basically the same information required for image-ray migration. This information consists of the local subsurface velocity distribution given as a function of depth (or two-way image time). Traveltimes of selected horizons are not required.

Figure 8-3 shows the surface of the earth. The half-space above it is the *time domain* for time-migrated data. Below the surface is the *depth domain*. It includes initially no more than the available local subsurface velocity distribu-

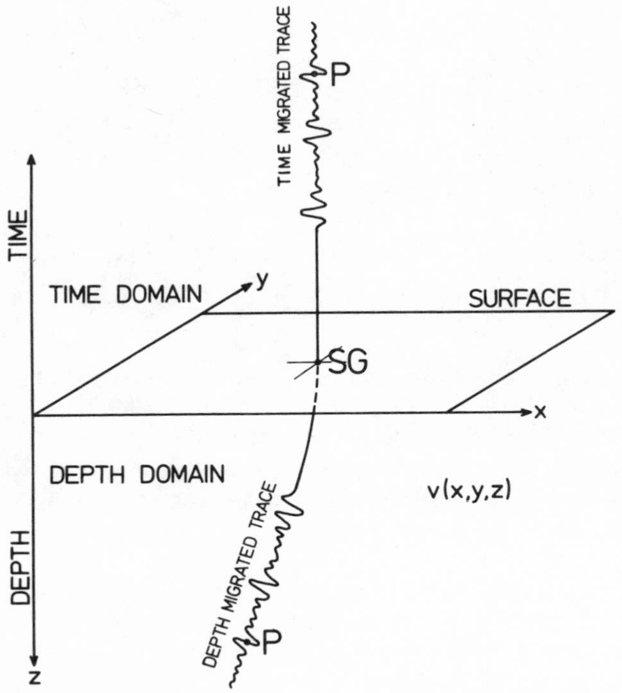

FIG. 8-3. Plot showing time-migrated trace in the time domain and depth-migrated trace (plotted along an image ray) in the depth domain.

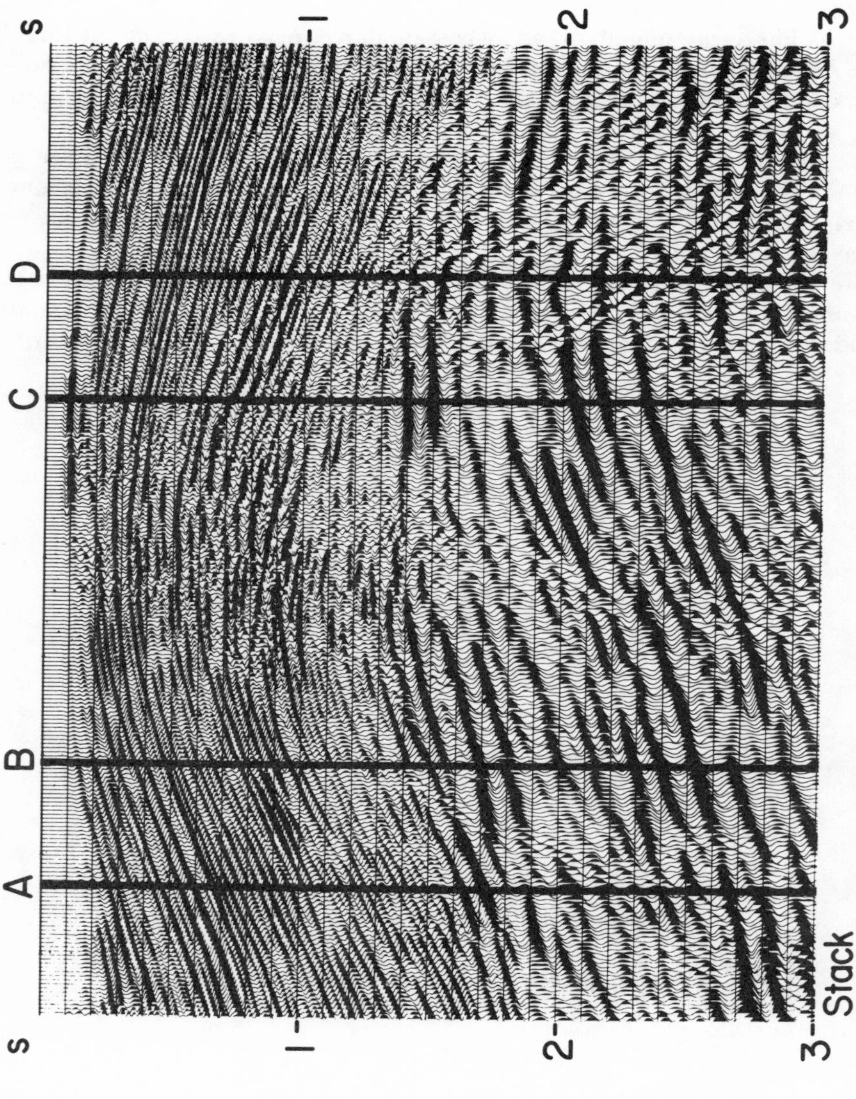

FIG. 8-4a. CDP stack section (courtesy Western Geophysical Co.).

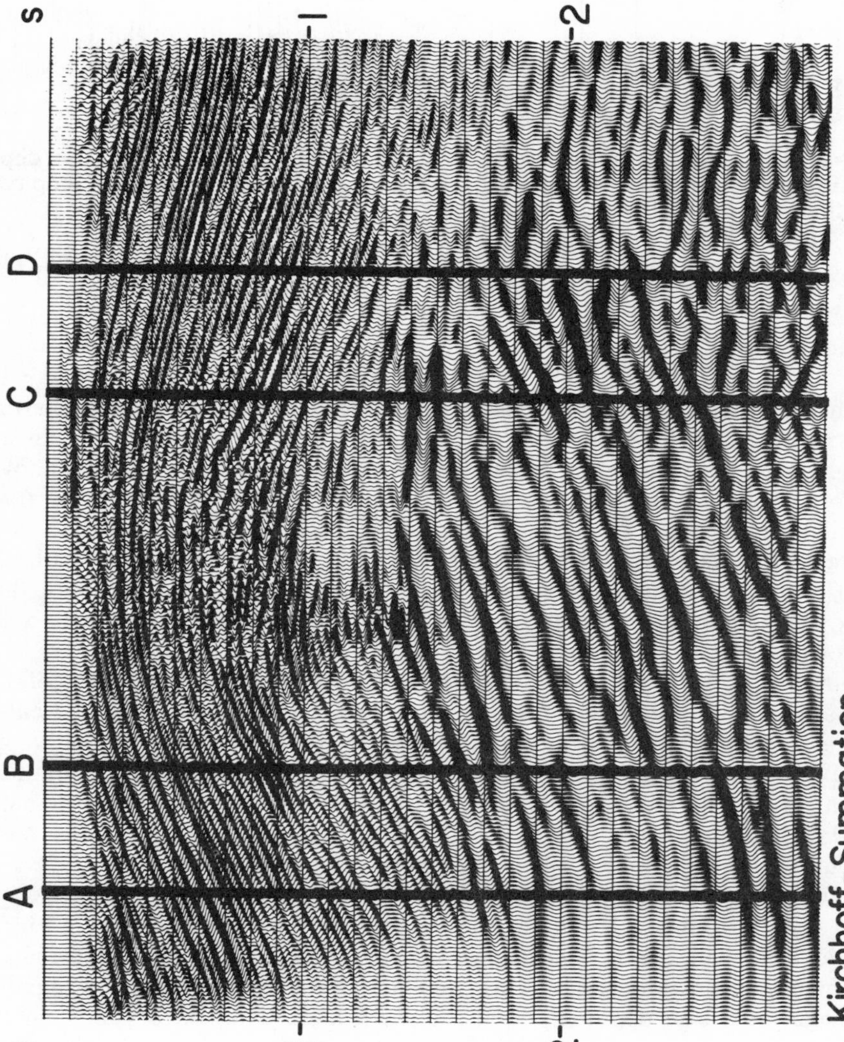

FIG. 8-4b. Time-migrated section (courtesy Western Geophysical Co.).

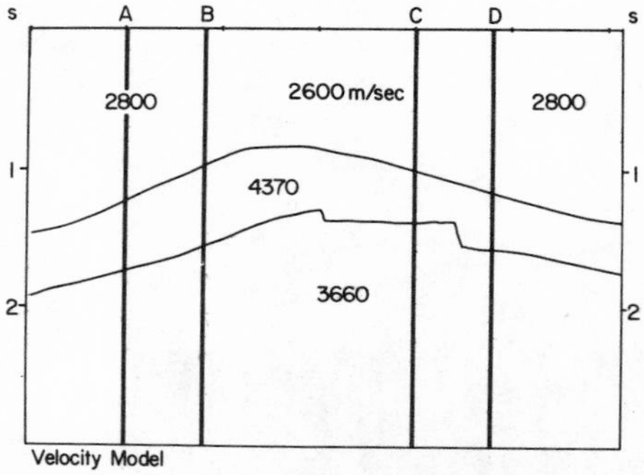

FIG. 8-4c. Local subsurface velocity distribution as a function of two-way image time for the CDP-stacked section (courtesy Western Geophysical Co.).

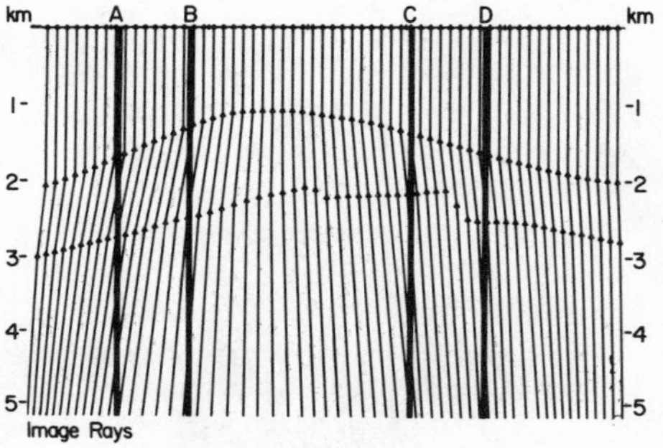

FIG. 8-4d. Image rays in the depth domain (courtesy Western Geophysical Co.).

tion. These velocity data are all that is needed to construct image rays starting at arbitrary surface locations. All time-migrated data of the time domain are depth-migrated into the depth domain in the following way. As shown above, the information contained at two-way image time T on a time-migrated trace positioned at SG stems from a scattering subsurface location D. The shortest traveltime from D to the earth's surface is $T/2$; this is associated with the (image) raypath that emerges at SG. Depth migration involves no more than moving the amplitude appearing at time T on the time-migrated trace to the depth location D.

Amplitudes of reflections on a time-migrated section are a measure of the energy scattered by the diffracting interface. The true *depth image* of the entire time-migrated trace at SG is indicated in Figure 8-3 in the form of the wiggly trace plotted along the image ray. Because this kind of display is of little practical value, in practice we would resample the data values placed in the depth domain. Then the displayed depth section will be represented by amplitudes on a rectangular grid, as is done for the time-domain data.

The following example illustrates some results. Figure 8-4a shows a CDP-stacked section from which the Kirchhoff summation migration, shown in Figure 8-4b, was obtained. Figure 8-4c shows the assumed local velocity distribution in the (migrated) time domain, and Figure 8-4d shows the depth domain together with image rays. The depth-migrated section (Figure 8-4e) differs most in the lower left corner from the conventional depth section (Figure 8-4f) which was obtained by vertically scaling the Kirchhoff summation section into depth.

As has been recognized previously, image rays may cross one another at depth, thus making a given subsurface location appear in two or more places on a time-migrated section. In such cases, diffraction time surfaces have more than one apex. Depth migration often can resolve this ambiguity and, can recover a true image of the earth properly scaled to depth and lateral coordinates.

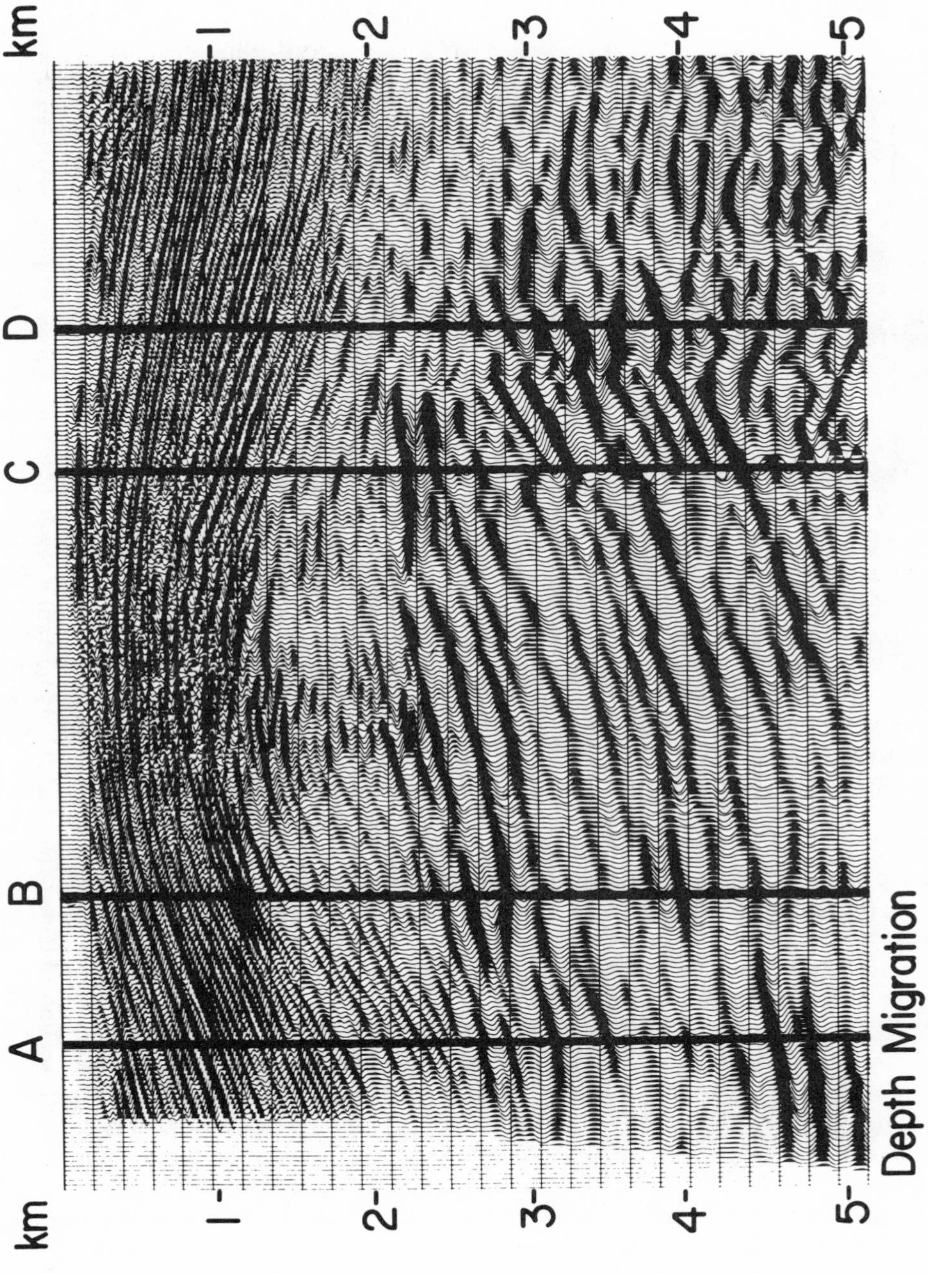

FIG. 8-4e. Depth-migrated section (courtesy Western Geophysical Co.).

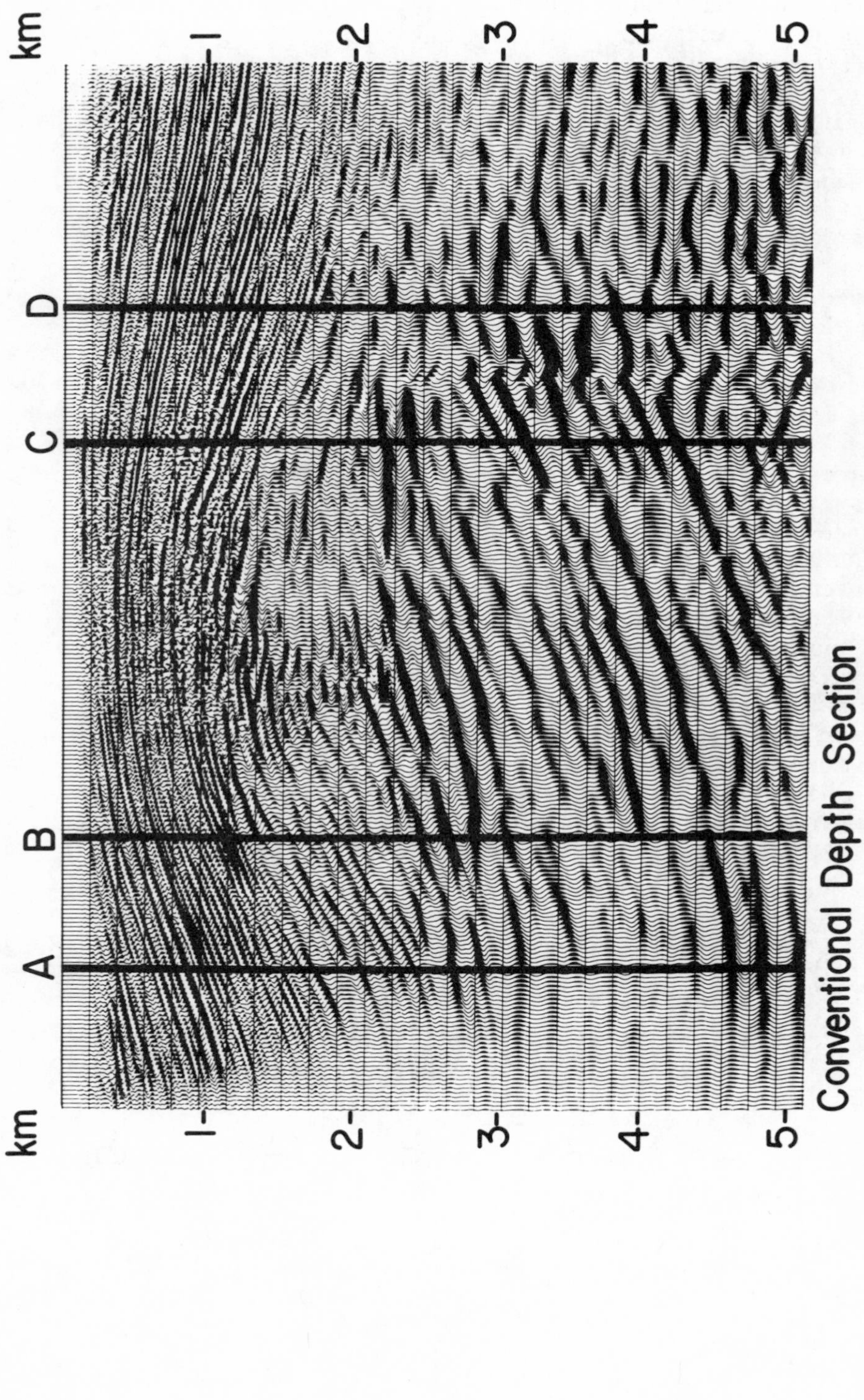

FIG. 8-4f. Depth section obtained from Kirchhoff migration result by vertical trace scaling (courtesy Western Geophysical Co.).

8.3.1 Summary

If subsurface reflectors are covered with a continuum of hypothetical sources exploding at time zero, then so-called *normal wavefronts* move up to the earth surface. Two-way normal reflection time maps describe twice the time at which such normal waves are recorded. That is, they describe the arrival times for normal waves that travel at a speed of half the actual medium velocity.

If subsurface velocities are known, one can effect a TDM by computationally propagating each normal wavefront backward from the surface of the earth to the interface where it originated. In addition, with knowledge of a second-order approximation to the two-way normal time map at a surface point where a normal ray emerges, one can compute the true interface curvature at the lower end of each normal ray.

If the earth's surface is covered by a continuum of sources exploding at time zero, a so-called *image wavefront* moves down into the earth. Two-way image time functions describe twice the time at which the image wavefront would be recorded when arriving at a particular interface.

Knowledge of the two-way image time at a surface location and knowledge of the subsurface velocity distribution allows one to perform *image-ray migration* or *depth migration*. These are just two of many possible TDM schemes. Furthermore, with a second-order approximation of the two-way image time function at a point on the surface of the earth, one can also compute true interface curvature at the lower end of the image ray. Interface curvatures are thus a new model parameter that can be accounted for in TDM algorithms.

Conventional TDM methods are based essentially on the concept of normal rays. With little difficulty, however, most existing computer programs can be easily modified to be used on time-migrated sections. Downward normal ray tracing need only be replaced by downward image ray tracing. These two types of rays, of course, obey the same laws of ray theory. Image ray tracing, however, is inherently more simple because the starting direction (i.e., vertical) is the same for all reflectors and all surface locations.

As we shall see in the next chapter, the ability to compute interface curvatures directly from reflection time surfaces is of particular help when computing interval velocities from traveltime measurements.

Depth migration is a process in which *all* the amplitude information in a time-migrated section is mapped to depth. Image ray modeling provides a simple and inexpensive way to determine whether seismic depth migration is necessary for a particular time-migrated section. Specifically, wherever refraction of image rays away from vertical is minimal, time-migrated data can be adequately converted to depth by simple vertical scaling.

9 Computation of interval velocity

We will now derive recursive algorithms for the computation of interval velocities using observed values of either NMO velocity V_{NMO} or SAM velocity V_{SAM}. Our initial considerations are confined to the purely mathematical aspects of solving this inverse traveltime problem. In later chapters, the more practical aspects will be discussed.

9.1 Interval velocities from NMO velocities

We shall assume that the local subsurface velocity distribution can be represented exactly by the layered model of Figure 6-4 and that the following quantities, (1) V_{NMO} within an arbitrary CDP profile, (2) two-way normal reflection time, and (3) lateral time gradient of normal reflection time, can be obtained by CDP techniques for all primary reflections resulting from reflecting first-order velocity boundaries at a given CDP point. We further assume that (below a given CDP point) the solution for the velocity model down to the $(N-1)$th interface has already been obtained.

We now show how interval velocity v_N can be obtained in a straightforward recursive fashion. Prior to describing the algorithm for the general 3-D model of Figure 6-4, let us discuss solutions for less complex 1- and 2-D models.

To start, we review the basic and familiar Dix algorithm. We shall give it a new interpretation, explaining it in terms of the particular philosophy of inversion employed throughout this monograph. This philosophy consists of recovering a recorded hypothetical wavefront from observed traveltimes and having it shrink back along a selected ray toward a hypothetical source. The particular hypothetical wavefront used in this section will be that of the NIP wave.

9.1.1 Dix method

The Dix algorithm is based on the model of Figure 4-8. The CDP reflection time curve for the uppermost reflector is

$$t_1^2(x) = t_1^2(0) + x^2/v_1^2 = L^2/v_1^2, \tag{9.1}$$

where L is the distance along the reflected raypath from source to reflector to geophone. For the uppermost reflector, we have $V_{\mathrm{NMO}} = V_{\mathrm{RMS}} = v_1$ so that v_1 is immediately available from the hyperbola (9.1). If the RMS velocities and two-way normal times are known to the $(N-1)$th and Nth reflectors, one can easily obtain v_N with Dix's formula

$$v_N = \left[\frac{V_{\mathrm{RMS},N}^2\, t_N(0) - V_{\mathrm{RMS},N-1}^2\, t_{N-1}(0)}{t_N(0) - t_{N-1}(0)}\right]^{1/2}. \tag{9.2}$$

This equation results immediately from the familiar expression for V_{RMS} given in equation (4.18). $V_{RMS,N}$ is the RMS velocity for the Nth reflective velocity boundary, and t_N (0) is the two-way (or one-way) time to the same reflector (Figure 4-8).

Once the velocity v_N is determined, we can immediately compute the depth to the Nth reflector, $d_N = d_{N-1} + \frac{1}{2} v_N \; [t_N(0) - t_{N-1}(0)]$, and the average velocity,

$$V_{A,N} = [1/t_N(0)] \sum_{k=1}^{N} v_k \; [t_k(0) - t_{k-1}(0)] .$$

If all layer boundaries have common dip and strike and the direction of the CDP profile coincides with the dip direction, then all normal rays lie in the same vertical seismic plane and emerge with the angle β_0; equation (9.2) is then replaced by

$$v_N = \cos \beta_0 \left[\frac{V^2_{NMO,N} \, t_N(0) - V^2_{NMO,N-1} \, t_{N-1}(0)}{t_N(0) - t_{N-1}(0)} \right]^{1/2} . \tag{9.3}$$

Using formula (4.30), one can obtain $\cos \beta_0$ from the common slope s_x of the two-way normal reflections times $t_\nu(0)(1 \leq \nu \leq N)$ along the seismic line

$$\cos \beta_0 = \left(1 - \frac{v_1^2}{4} s_x^2 \right)^{1/2} .$$

All quantities in equation (9.3) are thus available from seismic survey data.

Equation (9.3) is sometimes called the *dip-corrected* Dix formula. Its applicability is restricted to the very limited model described above. The accuracy to which v_N can be obtained from equation (9.3) depends only on the accuracy to which V_{NMO} and the two-way normal time to the upper and lower boundary of the Nth velocity layer are obtained. Inaccuracies with respect to other, shallower interfaces do not influence the accuracy of v_N. This interesting result is no longer valid for solutions involving more complex models. There the accuracy with which v_N can be obtained depends upon the accuracy of traveltime measurements of reflections related to all layer boundaries in the overburden.

The simplicity of equation (9.2) makes it appealing. It is even more attractive when one views it as resulting from a downward continuation process consistent with our general inversion philosophy. As indicated earlier, one can associate the CDP ray family of Figure 4-8 with a hypothetical NIP wavefront that originates at the NIP and emerges at the CDP. This association still holds true for more complex velocity models so long as offsets are sufficiently small (see Appendix D). The radius of curvature of the emerging NIP wavefront originating at O_N (=NIP) is

$$R_{0,N} = \frac{1}{2v_1} (v_1^2 \Delta t_1 + v_2^2 \Delta t_2 + \ldots, + v_N^2 \Delta t_N). \tag{9.4}$$

$R_{0,N}$ relates to $V_{NMO,N}$ [see equation (6.9)] as follows:

$$V^2_{\mathrm{NMO},N} = V^2_{\mathrm{RMS},N} = \frac{2v_1}{t_N(0)} R_{0,N}, \tag{9.5}$$

and

$$t_N(0) = \sum_{i=1}^{N} \Delta t_i, \tag{9.6}$$

where Δt_i is the two-way time in the ith layer.

Using equation (9.5), one can recover $R_{0,N}$ from $V_{\mathrm{RMS,N}}$, $t_N(0)$, and v_1. Provided the quantities v_i, d_i ($i = 1, \ldots, N-1$) are known, one can let the NIP wavefront shrink backward (toward its source at 0_N) as far down as to the lower side of 0_{N-1}. This reversal of wave propagation agrees with reversing the recursion which led to equation (9.4).

If, for instance, $R_{0,N}$ is the radius of the emerging NIP wavefront that originates in 0_N at the Nth interface, then

$$(R_{0,N} - v_1 \Delta t_1/2) \frac{v_1}{v_2},$$

$$\left[(R_{0,N} - v_1 \Delta t_1/2) \frac{v_1}{v_2} - v_2 \Delta t_2/2 \right] \frac{v_2}{v_3},$$

etc., are the radii of the same hypothetical NIP wavefront at the lower side of the first and second interface, etc.

By the time the lower side of 0_{N-1} is reached by the shrinking NIP wavefront, one is left with the following condition

$$v_N \Delta t_N/2 = \left[v_1 R_{0,N} - \frac{1}{2} \left(v_1^2 \Delta t_1 + v_2^2 \Delta t_2 + \ldots, + v_{N-1}^2 \Delta t_{N-1} \right) \right] / v_N,$$

which is obviously equal to

$$v_N^2 = \frac{2v_1 R_{0,N} - 2v_1 R_{0,N-1}}{\Delta t_N}, \tag{9.7}$$

where $R_{0,N-1}$ is the radius of curvature of the emerging NIP wavefront whose source is at 0_{N-1}. Equation (9.2) is obtained by simple substitution of equation (9.5) into equation (9.7).

The idea of downward propagating the NIP wavefront and reversing the algorithm for curvature of the NIP wavefront with respect to the normal ray is the key to recovering all subsurface velocities. To view the computation of interval velocities as a downward continuation process may, in the light of the simple Dix formula, appear a bit cumbersome for the case of plane parallel layers as treated above. It is, however, the most appealing approach when one considers more complex subsurface velocity models.

Interval velocities.—The real earth is generally far more finely stratified than can practically be considered for computation. Recognizing that we cannot hope to identify reflections from all possible interfaces, let us briefly look a little closer at the quantity we call interval velocity. When assuming a system of plane parallel horizontal isovelocity layers we can, at most, avail ourselves of the RMS velocities and two-way normal times of dominant reflectors that make up the upper and lower boundary of a *system* of say, M "micro-isovelocity" layers. If the upper interface index is i and the lower index $i + M$, then the "interval velocity" obtained from Dix's formula for the system of M micro-isovelocity layers is

$$
\begin{aligned}
V^2_{M,\,\mathrm{DIX}} &= \frac{V^2_{\mathrm{RMS},i+M}\, t_{i+M}(0) - V^2_{\mathrm{RMS},i}\, t_i(0)}{t_{i+M}(0) - t_i(0)}, \\
&= \frac{1}{t_{i+M}(0) - t_i(0)} \sum_{k=i+1}^{i+M} v_k^2\, [t_k(0) - t_{k-1}(0)]\,.
\end{aligned}
\tag{9.7a}
$$

The quantity $V_{M,\mathrm{DIX}}$ is the *RMS velocity of the interval.* It is larger than the interval velocity that would be given in form of the *average velocity* pertaining to the same layer boundaries.

Suppose average velocities are available from a well shot. If $V_{A,i}$ and $V_{A,i+M}$ are the average velocities (measured from the earth surface) to the layer boundaries under consideration, then one form of interval velocity could be obtained as follows:

$$
\begin{aligned}
V_{M,A} &= \frac{V_{A,\,i+M}\, t_{i+M}(0) - V_{A,i}\, t_i(0)}{t_{i+M}(0) - t_i(0)}, \\
&= \frac{1}{t_{i+M}(0) - t_i(0)} \sum_{k=i+1}^{i+M} v_k\, [t_k(0) - t_{k-1}(0)]\,.
\end{aligned}
\tag{9.7b}
$$

This is the *average velocity* of the total interval. If all M isovelocity layers in the interval had the same velocity, then the interval velocity as defined by equation (9.7b) would be identical to that given by equation (9.7a). However, as we see here, if the presumption of constant velocity within the interval is not fulfilled, interval velocities are quantities that depend on the method by which they are calculated.

In practice, as CVL logs reveal, the local velocity function typically is rather wiggly within any interpreted "macro-interval." We, therefore, must be aware that the *RMS velocity of a macro-interval* as determined here (and then called interval velocity) differs from the average velocity for this macro-interval. Referring to our N (so-called) macro-isovelocity layers, we find in practice that these differences within layers are small compared to the difference between RMS velocity and average velocity for the combination of all N layers.

When a CVL is available somewhere in the neighborhood of the reflection survey, the difference between $V_{M,\mathrm{DIX}}$ (9.7a) and $V_{M,A}$ (9.7b) can be computed and serve as an estimate for quantities derived from surface seismic measurements. It has to be kept in mind, however, that strongly fluctuating velocity behavior within thin depth intervals gives rise to yet a different type of discrepancy between RMS velocity and average velocity. In particular, where those intervals are small as compared to the wavelength of our signals, anisotropy results wherein the velocity in the direction perpendicular to stratification is smaller than the log-computed average velocity (see Helbig, 1965; Uhrig and van Melle, 1955). Since NMO velocity is influenced by the speed of propagation in the horizontal as well as the vertical direction (Thomas and Lucas, 1977), the estimated RMS velocity will be likewise influenced.

9.1.2 2-D algorithms

The problem of recovering the 2-D plane isovelocity layer model of Figure 6-2 from CDP surface measurements has been solved by various authors (Larner and Rooney, 1972; Everett, 1974; Késmarky, 1977) in different ways. Késmarky's method is approximate and does not require any ray tracing at all. The traveltime inversion methods described by these authors are already fairly complex as compared with the simple Dix algorithm. They differ from the procedure discussed here.

Provided the layer velocities v_i ($i = 1, \ldots, N-1$) are known and the upper $N-1$ velocity interfaces are already constructed, then the following steps lead to the determination of v_N and the positioning of the Nth reflector. They are in agreement with our proposed inversion philosophy:

(1) Find the emergence angle β_0 from the gradient of the normal reflection times and from v_1 at the CDP (see formula 4.30).

(2) Recover the radius of curvature $R_{0,N}$ of the emerging NIP wavefront from v_1, $V_{\mathrm{NMO,N}}$, $t_N(0)$, and β_0 at the CDP with formula (6.9).

(3) Reverse the recursion of the NIP wavefront. This inverse recursion pertains to the normal ray (at 0_N), which can be traced from the CDP as far as 0_{N-1}. The first few steps are

$$R_{I,1} = -R_{0,N} + s_1,$$

$$R_{T,1} = \frac{v_1 \cos^2 \beta_1}{v_2 \cos^2 \alpha_1} R_{I,1},$$

and

$$R_{I,2} = R_{T,1} + s_2.$$

The radii of the shrinking NIP wavefront are here chosen to be of opposite sign from the radii of the upward expanding NIP wavefront considered in the forward problem. With this choice, we can measure traveltimes and distances positive when tracing the normal ray down. $R_{I,1}$ is the radius of curvature on the upper and $R_{T,1}$ on the lower side of point 0_1.

The final two recursive steps are

$$R_{T,N-1} = \frac{v_{N-1} \cos^2 \beta_{N-1}}{v_N \cos^2 \alpha_{N-1}} R_{I,N-1},$$

and

$$R_{I,N} = 0 = R_{T,N-1} + v_N \Delta t_N / 2.$$

They provide the following equation

$$-v_N \Delta t_N / 2 = \frac{v_{N-1} \cos^2 \beta_{N-1}}{v_N \cos^2 \alpha_{N-1}} R_{I,N-1}.$$

By combining this relationship with the two conditions,

$$\Delta t_N = t_N(0) - \sum_{i=1}^{N-1} \Delta t_i, \tag{9.8}$$

and

$$\frac{\sin \alpha_{N-1}}{v_{N-1}} = \frac{\sin \beta_{N-1}}{v_N}, \tag{9.9}$$

we obtain the following quadratic equation for v_N

$$v_N^2 \left(\frac{\Delta t_N}{2} - \frac{\sin^2 \alpha_{N-1}}{v_{N-1} \cos^2 \alpha_{N-1}} R_{I,N-1} \right) + \frac{v_{N-1}}{\cos^2 \alpha_{N-1}} R_{I,N-1} = 0. \tag{9.10}$$

The positive real root for v_N provides the desired solution. With it, we obtain β_{N-1} immediately from equation (9.9).

Δt_i is the two-way time in layer i along the normal ray. The quantities β_{N-1}, v_N, and Δt_N are then sufficient to determine the position of 0_N. The Nth interface is then completely specified when we recognize that it must be perpendicular to the normal ray at 0_N.

Rather than using the above inverse traveltime recursion, we could as well use the analytical expression for V_{NMO}, substitute all known parameters into it that are obtained along the normal ray, and solve the equation for v_N in conjunction with equations (9.8) and (9.9). This procedure has been described in Hubral (1974). Adapting the latter approach to the more complex cases discussed next, however, would be a difficult undertaking. We will, therefore, follow the recursive approach to solving the inverse traveltime problem and disregard the explicit approach.

The algorithm to recover 2-D curved isovelocity layers (Figure 6-3) from CDP surface measurements is little different from the one for plane dipping interfaces. Provided v_i ($i = 1, \ldots, N-1$) and the upper $N-1$ interfaces are already obtained, then v_N and the location 0_N are recovered by accompanying the downward tracing of the normal ray from the CDP to 0_{N-1} with the reverse recursion that led to equation (6.11). This recursion is as follows

$$\begin{aligned}
R_{I,1} &= -R_{0,N} + s_1, \\
R_{T,1}^{-1} &= \frac{v_2 \cos^2 \alpha_1}{v_1 \cos^2 \beta_1} R_{I,1}^{-1} + \frac{1}{\cos^2 \beta_1} \left(\frac{v_2}{v_1} \cos \alpha_1 - \cos \beta_1 \right) R_{F,1}^{-1}, \\
&\cdots\cdots\cdots\cdots \\
R_{T,N-1}^{-1} &= \frac{v_N \cos^2 \alpha_{N-1}}{v_{N-1} \cos^2 \beta_{N-1}} R_{I,N-1}^{-1} \\
&\quad + \frac{1}{\cos^2 \beta_{N-1}} \left(\frac{v_N}{v_{N-1}} \cos \alpha_{N-1} - \cos \beta_{N-1} \right) R_{F,N-1}^{-1}, \\
R_{I,N} &= 0 = R_{T,N-1} + v_N \Delta t_N / 2.
\end{aligned} \tag{9.11}$$

Again, the last two equations can be combined with equations (9.8) and (9.9), this time yielding a biquadratic equation of fourth degree for v_N.

In addition, the curvature $R_{F,N}^{-1}$ of the Nth interface at 0_N can be obtained

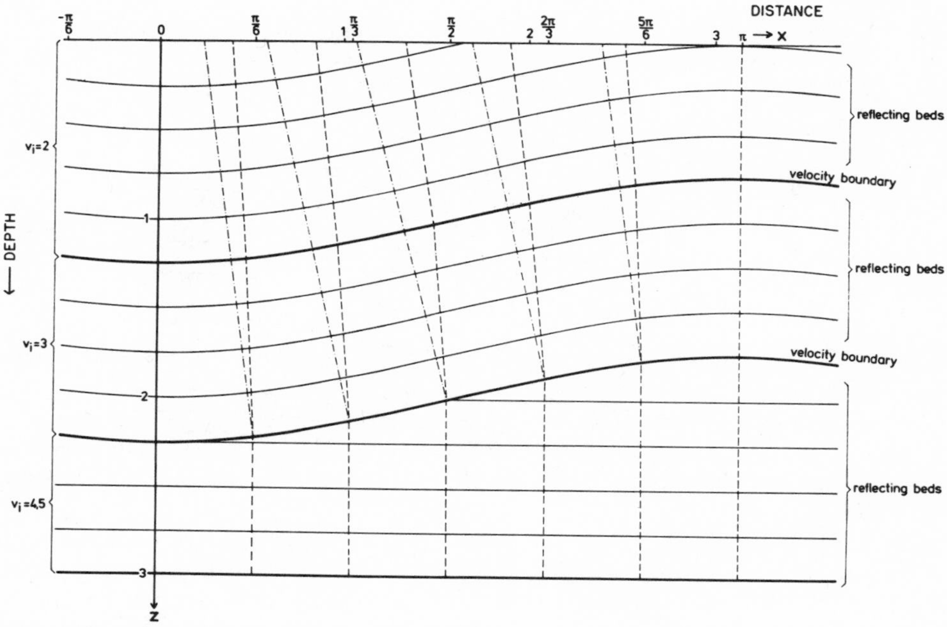

FIG. 9-1. 2-D curved layer model with additional reflecting interfaces and normal rays (dashed lines) to the second and third velocity boundaries.

directly using the normal ray migration algorithm discussed in section 8.1. This approach is considerably more efficient than the more conventional one requiring determination of the depth horizon of the entire Nth interface in the form of many NIPs. As discussed earlier, when dealing with NIP waves, we recognized that the curvatures $R_{F,j}^{-1}$ ($j = 1, 2, \ldots, N-1$) in equation (9.11) are available from interpolation of solutions obtained for the respective interfaces at nearby CDPs.

Example.—The following example reveals the influence of dip and curvature on V_{NMO} and the inadequacy of the Dix algorithm for the computation of interval velocities in the presence of curved reflecting velocity boundaries.

Figure 9-1 shows a subsurface model consisting of three homogeneous layers having constant velocities 2, 3, and 4.5 km/sec, respectively. The velocity boundaries are presented by $z = 1 + (1/4) \cos x$ and $z = 2 + (1/4) \cos x$, where z is the vertical and x is the horizontal coordinate; both are measured in kilometers. Some normal rays to the second and third velocity boundary are indicated in the figure. Each of the three major layers is further subdivided by additional reflecting horizons that parallel the lower-velocity boundaries of the respective layers. The model provides a rough approximation to a Central European halokinetic area, the first layer being Tertiary, the second Mesozoic, and the third Permian salt.

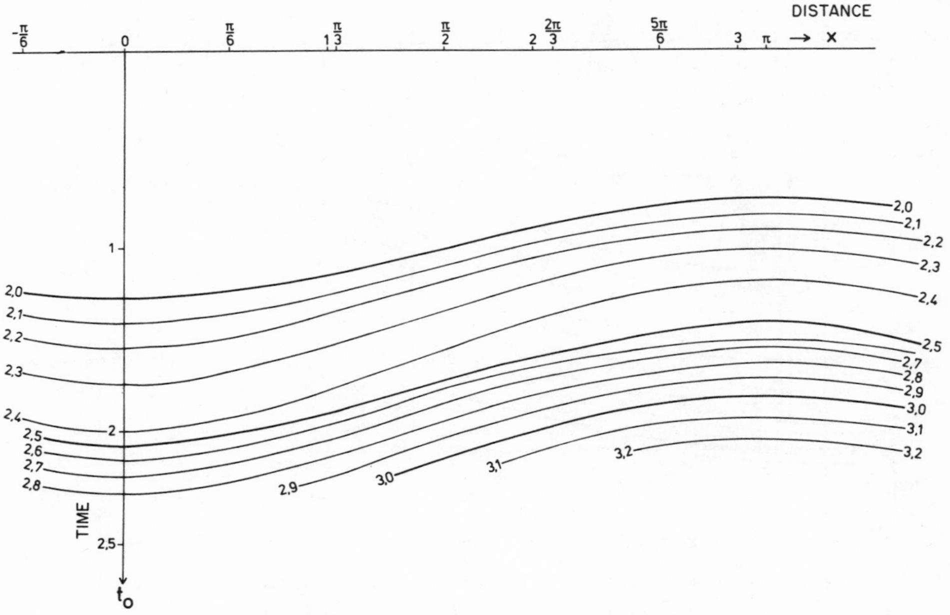

FIG. 9-2. Contours of RMS velocity (km/sec).

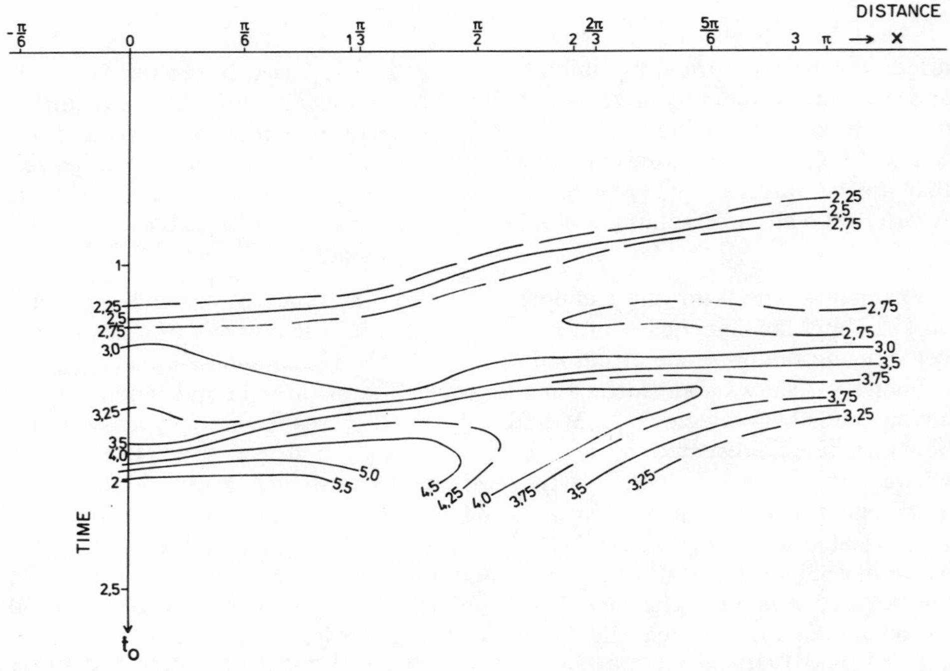

FIG. 9-3. Contours of NMO velocity (km/sec).

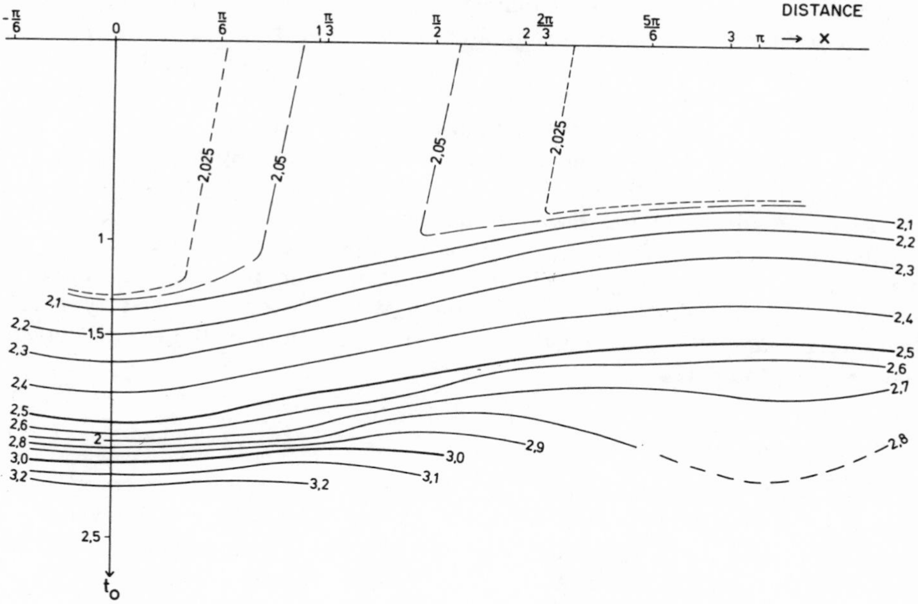

FIG. 9-4. Contours of local velocities (km/sec) computed by the dip-corrected Dix formula using NMO velocities and two-way normal times.

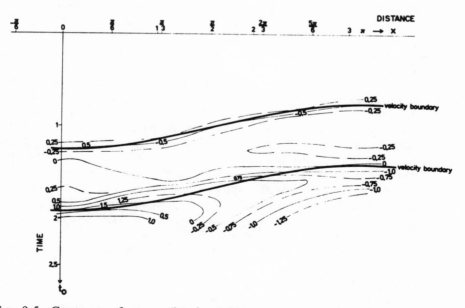

FIG. 9-5. Contours of errors (km/sec) between true local velocities and those computed by the dip-corrected Dix formula.

Figure 9-2 shows the RMS velocities of the reflecting beds computed along the rays of Figure 9-1 and represented as contours in the normal time t_0 versus distance x coordinate system; Figure 9-3 shows the contoured NMO velocities. These contours were established for a set of CDPs and numerous reflecting subsurface points. They are presented in the same manner.

The true local velocities and reflecting horizons in depth, of course, can be recovered exactly with the above 2-D curved layer algorithm. If, however, one used the dip-corrected Dix formula (9.3) to establish the local velocity distribution, then the derived interval velocities would be quite erroneous and unacceptable over many portions of the section. The resulting velocities are shown in Figure 9-4. (We computed them with a constant two-way normal time interval $\Delta t = 0.2$ sec.) Figure 9-5 provides contours of the errors between true and Dix-computed velocities.

9.1.3 3-D algorithm

The NMO velocity for the Nth primary CDP reflection observed within some specified profile (Figure 6-4) can be obtained by relating it to the wavefront curvature matrix $\mathbf{A}_0$ of the emerging hypothetical NIP wavefront at the CDP. According to equation (6.14), $\mathbf{A}_0$ is recoverable from NMO velocities observed in three different directions through the CDP.

Suppose the upper $N-1$ layer shapes and velocities have already been recovered, and $\mathbf{A}_0$ is known for the Nth primary CDP reflection at the CDP. Then shrinking the NIP wavefront along the normal ray back into its hypothetical source at 0_N means reversing the matrix recursion of the type that led to expression (6.16) in the three-layer case.

The direction of the emerging normal ray can be found from v_1 and the normal time gradient (see equation 4.30). Provided the Nth velocity layer is not as yet known, the normal ray from CDP can be traced only as far down as 0_{N-1}. When this ray tracing is accompanied by the reverse of the recursion that leads to $\mathbf{A}_0$ in the forward problem, one again obtains some simple equations for the determination of v_N.

Though this procedure is in agreement with our general inversion philosophy, one might conclude (wrongly) that three CDP profiles are needed through a CDP in order to solve the 3-D inverse traveltime problem directly. Surprisingly, this is not so, as will be explained in some detail on page 130 below.

The minimum information required for solution, in fact, consists of the NMO velocity from just *one* CDP profile (Figure 6-4), the two-way normal time, and the gradient of the Nth normal time map at CDP. Surprisingly, then, one can recover the 3-D shape of the reflecting velocity boundaries from only observations of NMO velocities within conventionally shot straight seismic lines plus the information available in contoured maps of normal primary reflection times. The seismic lines need not necessarily cross one another. NMO velocities in crossing CDP profiles thus provide redundant information not strictly necessary in solving for interval velocities.

Let us now see how v_N can be obtained.

Suppose, for the moment, that $\mathbf{A}_0$ (in the $[x, y, z]$ system of the emerging normal ray whose z-axis points up) is completely available at CDP. Knowing the upper $N-1$ velocity layers and the direction of the emerging normal ray to the Nth reflector at CDP, we can trace the normal ray down to 0_{N-1} (Figure 6-4). We accompany the downward tracing of the normal ray with the following recursion (see section 4.3.5) pertaining to the moving frame whose z-axis points downward.

$$\mathbf{A}_{I,1}^{-1} = -\mathbf{A}_0^{-1} + s_1\,\mathbf{I}, \tag{9.12a}$$

$$\mathbf{A}_{T,1} = \mathbf{D}_1^{-1}\left(\frac{v_2}{v_1}\,\mathbf{S}_1\,\mathbf{A}_{I,1}\,\mathbf{S}_1 + \rho_1\,\mathbf{S}_{T,1}^{-1}\,\mathbf{B}_1\,\mathbf{S}_{T,1}^{-1}\right)\mathbf{D}_1, \tag{9.12b}$$

. . . .

$$\mathbf{A}_{I,\nu}^{-1} = \mathbf{A}_{T,\nu-1}^{-1} + s_\nu\,\mathbf{I}, \tag{9.12c}$$

$$\mathbf{A}_{T,\nu} = \mathbf{D}_\nu^{-1}\left[\frac{v_{\nu+1}}{v_\nu}\,\mathbf{S}_\nu\,\mathbf{A}_{I,\nu}\,\mathbf{S}_\nu + \rho_\nu\,\mathbf{S}_{T,\nu}^{-1}\,\mathbf{B}_\nu\,\mathbf{S}_{T,\nu}^{-1}\right]\mathbf{D}_\nu, \tag{9.12d}$$

. . . .

$$\mathbf{A}_{I,N-1}^{-1} = \mathbf{A}_{T,N-2}^{-1} + s_{N-1}\,\mathbf{I}, \tag{9.12e}$$

$$\mathbf{A}_{T,N-1} = \left(\frac{v_N}{v_{N-1}}\,\mathbf{S}_{N-1}\,\mathbf{A}_{I,N-1}\,\mathbf{S}_{N-1} + \rho_{N-1}\,\mathbf{S}_{T,N-1}^{-1}\,\mathbf{B}_{N-1}\,\mathbf{S}_{T,N-1}^{-1}\right), \tag{9.12f}$$

$$\mathbf{N} = \mathbf{A}_{T,N-1}^{-1} + s_N\,\mathbf{I} = \begin{bmatrix} 0 & 0 \\ 0 & 0 \end{bmatrix}. \tag{9.12g}$$

$\mathbf{S}_i$, $\mathbf{S}_{T,i}$, $\mathbf{D}_i$, and ρ_i are the quantities defined in section 4.3.5, and $\mathbf{B}_\nu$ ($\nu = 1, 2, \ldots, N-1$) have been obtained by interpolation along previously determined interfaces.

$\mathbf{A}_{I,\nu}$ is the wavefront curvature matrix on the upper and $\mathbf{A}_{T,\nu}$ on the lower side of the νth velocity interface. In equation (9.12a), we have used the negative sign in $-\mathbf{A}_0^{-1}$ for the shrinking NIP wavefront so that traveltimes and distances downward along the normal ray will be measured positive. There is no need for a matrix $\mathbf{D}_{N-1}$ in equation (9.12f) (i.e., we may assume $\mathbf{D}_{N-1} = \mathbf{I}$) because any plane perpendicular to the Nth interface at 0_N may be chosen as the plane of incidence for 0_N. Thus, we can keep the plane of incidence at 0_{N-1} as the plane of incidence for 0_N.

Equation (9.12g) expresses the fact that the radius matrix of the NIP wavefront shrinks back to zero at 0_N. The last two of equations (9.12) include the unknowns v_N, β_{N-1}, and $s_N = v_N\,\Delta t_N/2$ which, in conjunction with the familiar conditions

$$\Delta t_N = t_N(0) - \sum_{i=1}^{N-1} \Delta t_i \tag{9.13}$$

$$\frac{\sin \alpha_{N-1}}{v_{N-1}} = \frac{\sin \beta_{N-1}}{v_N}, \tag{9.14}$$

can be solved for v_N.

The approach just described shows how we can compute v_N from observations of the curvature matrix $\mathbf{A}_0$, the normal traveltime, and its gradient. In order to demonstrate that we need observations of V_{NMO} in only one direction (rather than the three directions for complete determination of $\mathbf{A}_0$), let us now approach the solution for v_N from yet another point of view.

Let us introduce two new matrices $\mathbf{H}_{I,\nu}$ and $\mathbf{H}_{T,\nu}$ by writing

$$\mathbf{A}_{I,\nu} = \prod_{\mu=\nu-1}^{1} (\mathbf{D}_{\mu}^{-1}\mathbf{S}_{\mu})\,\mathbf{H}_{I,\nu} \prod_{\mu=1}^{\nu-1} (\mathbf{S}_{\mu}\mathbf{D}_{\mu}) \tag{9.15a}$$

and

$$\mathbf{A}_{T,\nu} = \prod_{\mu=\nu}^{1} (\mathbf{D}_{\mu}^{-1}\mathbf{S}_{\mu})\,\mathbf{H}_{T,\nu} \prod_{\mu=1}^{\nu} (\mathbf{S}_{\mu}\mathbf{D}_{\mu}). \tag{9.15b}$$

Then matrix equations (9.12c) and (9.12d) can be written as

$$\mathbf{H}_{I,\nu}^{-1} = \mathbf{H}_{T,\nu-1}^{-1} + s_{\nu} \prod_{\mu=1}^{\nu-1} (\mathbf{S}_{\mu}\mathbf{D}_{\mu}) \prod_{\mu=\nu-1}^{1} (\mathbf{D}_{\mu}^{-1}\mathbf{S}_{\mu}), \tag{9.15c}$$

$$= \mathbf{H}_{T,\nu-1}^{-1} + \mathbf{P}_{\nu} \tag{9.15d}$$

$$\mathbf{H}_{T,\nu} = \frac{v_{\nu+1}}{v_{\nu}}\mathbf{H}_{I,\nu} + \rho_{\nu} \prod_{\mu=1}^{\nu} (\mathbf{S}_{\mu}^{-1}\mathbf{D}_{\mu})\,\mathbf{S}_{T,\nu}^{-1}\mathbf{B}_{\nu}\mathbf{S}_{T,\nu}^{-1} \prod_{\mu=\nu-1}^{1} (\mathbf{D}_{\mu}^{-1}\mathbf{S}_{\mu}^{-1}), \tag{9.15e}$$

$$= \frac{v_{\nu+1}}{v_{\nu}}\mathbf{H}_{I,\nu} + \mathbf{Q}_{\nu}, \tag{9.15f}$$

where we have used the definitions

$$\mathbf{P}_{\nu} \equiv s_{\nu} \prod_{\mu=1}^{\nu-1} (\mathbf{S}_{\mu}\mathbf{D}_{\mu}) \prod_{\mu=\nu-1}^{1} (\mathbf{D}_{\mu}^{-1}\mathbf{S}_{\mu}), \tag{9.15g}$$

and

$$\mathbf{Q}_{\nu} = \rho_{\nu} \prod_{\mu=1}^{\nu} (\mathbf{S}_{\mu}^{-1}\mathbf{D}_{\mu})\,\mathbf{S}_{T,\nu}^{-1}\mathbf{B}_{\nu}\mathbf{S}_{T,\nu}^{-1} \prod_{\mu=\nu-1}^{1} (\mathbf{D}_{\mu}^{-1}\mathbf{S}_{\mu}^{-1}). \tag{9.15h}$$

$\mathbf{H}_{I,\nu}$ and $\mathbf{H}_{T,\nu}$ thus satisfy the following recursion scheme,

$$\mathbf{H}_{I,\nu}^{-1} = \mathbf{H}_{T,\nu-1}^{-1} + \mathbf{P}_{\nu}, \tag{9.15d}$$

$$\mathbf{H}_{T,\nu} = \frac{v_{\nu+1}}{v_{\nu}}\mathbf{H}_{I,\nu} + \mathbf{Q}_{\nu}, \tag{9.15f}$$

with

$$\mathbf{H}_{T,0} = -\mathbf{A}_0.$$

If equation (9.15d) is substituted into equation (9.15f), $\mathbf{H}_{T,\nu}$ can be represented by a finite continued fraction of the kind provided by equation (6.16). The same is true for $\mathbf{H}_{I,\nu}$ if equation (9.15f) is substituted into equation (9.15d).

Continued fractions can be avoided by doubling the number of unknown matrices; i.e., by introducing the matrices $\mathbf{J}_{T,\nu}$ and $\mathbf{K}_{T,\nu}$ as follows

$$\mathbf{H}_{T,\nu} = \mathbf{K}_{T,\nu}^{-1}\mathbf{J}_{T,\nu},$$

so that (9.15i)

$$\mathbf{H}_{T,\nu}^{-1} = \mathbf{J}_{T,\nu}^{-1}\mathbf{K}_{T,\nu},$$

whereby the initial matrices are given as

$$\mathbf{K}_{T,0} = \mathbf{I} \tag{9.15j}$$

and

$$\mathbf{J}_{T,0} = \mathbf{H}_{T,0} = -\mathbf{A}_0.$$

Equations (9.15d) and (9.15f) can then be satisfied by

$$\mathbf{J}_{T,\nu} = \mathbf{J}_{T,\nu-1}\left(\frac{v_{\nu+1}}{v_\nu}\mathbf{I} + \mathbf{P}_\nu\,\mathbf{Q}_\nu\right) + \mathbf{K}_{T,\nu-1}\,\mathbf{Q}_\nu,$$

and

$$\mathbf{K}_{T,\nu} = \mathbf{J}_{T,\nu-1}\,\mathbf{P}_\nu + \mathbf{K}_{T,\nu-1}, \tag{9.15k}$$

as can be easily verified.

By introducing the 2×4 matrix

$$\mathbf{L}_\nu = (\mathbf{J}_{T,\nu}, \mathbf{K}_{T,\nu}) \tag{9.15l}$$

and the 4×4 matrix

$$\mathbf{T}_\nu = \begin{bmatrix} \dfrac{v_{\nu+1}}{v_\nu}\mathbf{I} + \mathbf{P}_\nu\mathbf{Q}_\nu & \mathbf{P}_\nu \\ \mathbf{Q}_\nu & \mathbf{I} \end{bmatrix}, \tag{9.15m}$$

we can replace (9.15k) by

$$\mathbf{L}_\nu = \mathbf{L}_{\nu-1}\,\mathbf{T}_\nu. \tag{9.15n}$$

Carrying out the recursion, we obtain

$$\mathbf{L}_{N-1} = \mathbf{L}_0 \prod_{\nu=1}^{N-1} \mathbf{T}_\nu. \tag{9.15p}$$

Although this is a simple compact expression, it is unlikely to reduce the necessary computing work indicated by equations (9.15c) to (9.15h). Equation (9.15p) corresponds to equation (5.24) previously given by Krey (1976). If all three elements of the wavefront curvature matrix $\mathbf{A}_0$ are known, $\mathbf{A}_{I,N-1}$ can be computed from equations (9.15a, b, d, f, g, and h). If preferred, equations (9.15i) and (9.15l–9.15p) could be used instead.

Before completing our argument, let us detour briefly to give some detailed attention to the problem of computing the elements a_{11}°, a_{12}°, and a_{22}° of the $\mathbf{A}_0$ matrix from primary CDP traveltime measurements. $\mathbf{A}_0$ is completely determined by measurements of V_{NMO} in three different directions through a CDP point ($y_s = x_s \tan \phi_i$ with $i = 1, 2, 3$) provided the gradient $\nabla\, t_N(x,y)$ of the normal reflection times $t_N(x,y)$ is also known at the CDP.

In the $[x_s, y_s]$ system at the CDP on the surface of the earth (see Figure 6-4), the NMO time can be related to the NMO velocity as follows

$$\Delta t_{\text{NMO}}(x_s, y_s) = (a_{11}\,x_s^2 + 2a_{12}\,x_s y_s + a_{22}\,y_s^2)/v_1 \tag{9.16}$$

$$= \frac{2(x_s^2 + y_s^2)}{t_N(0,0)\,V_{\text{NMO}}^2(\phi_s)}, \tag{9.17}$$

whereby terms of higher than the second power in x_s and y_s are neglected again.

Thus, the knowledge of either Δt_{NMO} at three noncollinear (x_s, y_s) points or V_{NMO} for three different azimuths defined by (y_s/x_s) is sufficient for the computation of a_{11}, a_{12}, and a_{22}. How are these traveltime parameters then converted into the components of the wavefront curvature matrix $\mathbf{A}_0$?

To answer this question, let us consider the plane tangent to the NIP wavefront at point 0_0 $(x_s = y_s = z_s = 0)$ within the $[x_{f0}, y_{f0}, z_{f0}]$ system, the emerging moving $[x, y, z]$ system at CDP. This system results from the $[x_s, y_s, z_s]$ system by turning it by the angle β_0 around the $y_s = y_{f0}$-axis, where

$$\sin \beta_0 = \frac{v_1}{2}\left(\frac{\partial t_N}{\partial x_s}\right)_{x_s=y_s=0}. \tag{9.18}$$

Then the normal moveout time at point x_s, y_s on the surface of the earth ($z_s = 0$) is the same as that at point $x_{f0} = x_s \cos \beta_0$, $y_{f0} = y_s$ in the tangential plane $z_{f0} = 0$. Thus, again neglecting higher than second power terms, we can write

$$\begin{aligned} &a_{11} x_s^2 + 2a_{12} x_s y_s + a_{22} y_s^2 \\ &= a_{11}^{\circ} x_{f0}^2 + 2a_{12}^{\circ} x_{f0} y_{f0} + a_{22}^{\circ} y_{f0}^2, \end{aligned} \tag{9.19}$$

where

$$\begin{bmatrix} a_{11}^{\circ} & a_{12}^{\circ} \\ a_{12}^{\circ} & a_{22}^{\circ} \end{bmatrix} = \mathbf{A}_0. \tag{9.20}$$

The components of $\mathbf{A}_0$ consequently are obtained from

$$a_{11}^{\circ} = a_{11}/\cos^2 \beta_0; \; a_{12}^{\circ} = a_{12}/\cos \beta_0; \; a_{22}^{\circ} = a_{22}, \tag{9.21}$$

or

$$\mathbf{A}_0 = \mathbf{S}_{T,0}^{-1} \, \mathbf{A} \, \mathbf{S}_{T,0}^{-1},$$

with

$$\mathbf{S}_{T,0} = \begin{bmatrix} \cos \beta_0 & 0 \\ 0 & 1 \end{bmatrix},$$

and

$$\mathbf{A} = \begin{bmatrix} a_{11} & a_{12} \\ a_{12} & a_{22} \end{bmatrix}.$$

When carrying out the recursion from $\mathbf{A}_{I,N-1}$ to $\mathbf{A}_{I,N}$, the unknown velocity v_N enters into relations (9.15a), (9.15b), (9.15d), (9.15f), etc. Because the hypothetical NIP wavefront shrinks completely to the normal incidence point at interface N, we have

$$\mathbf{A}_{I,N}^{-1} = \mathbf{N} = \begin{bmatrix} 0 & 0 \\ 0 & 0 \end{bmatrix}, \tag{9.22}$$

or, by equation (9.15a),

$$\mathbf{H}_{I,N}^{-1} = \mathbf{N} = \begin{bmatrix} 0 & 0 \\ 0 & 0 \end{bmatrix}. \tag{9.23}$$

Thus, with the exception of $\mathbf{A}_0$, we need not deal directly with any of the $\mathbf{A}$ matrices. It suffices to consider the equations (9.15d) and (9.15f) together with definitions (9.15g) and (9.15h). Alternatively, we may apply equations (9.15n) and (9.15p) with definitions (9.15i) and (9.15j). In that case, according to (9.15i), equation (9.23) can be replaced by

$$\mathbf{K}_{T,N} = \mathbf{N} = \begin{bmatrix} 0 & 0 \\ 0 & 0 \end{bmatrix}. \tag{9.24}$$

So long as $\mathbf{A}_0$ is known completely, each of the matrix equations (9.22)–(9.24) comprises four scalar equations, three of which are independent. These three independent equations constitute an overdetermined system for the single unknown v_N.

If, in practice (e.g., for good noise-free reflection data) the four equations cannot be reconciled, our model may have been too simple. For example, the velocity of the Nth layer might be far from homogeneous or it might be strongly anisotropic.

As already indicated, the overdetermined set of equations points to the fact that we do not really need to measure NMO velocities (or normal moveout times) in three different crossing CDP profile directions. One profile direction is, in fact, sufficient.

In order to complete the proof, let us suppose that (along with v_N) a_{11}°, a_{12}°, and a_{22}° are unknown. Any of equations (9.22), (9.23), and (9.24) represent three independent equations for the unknown quantities a_{11}°, a_{12}°, a_{22}°, and v_N. If equation (9.24) is used, these equations are *linear* in a_{11}°, a_{12}°, a_{22}° as could be verified from equations (9.15j) to (9.15p).

A fourth linear equation for a_{11}°, a_{12}°, and a_{22}° is provided by equations (9.16), (9.17), and (9.21) as follows.

In Figure 6-4, we have

$$x_s = \frac{r}{2}\cos\phi_s, \quad y_s = \frac{r}{2}\sin\phi_s,$$

r being the shot-geophone distance or CDP offset in the CDP profile concerned. After dividing by r^2, equations (9.16) and (9.17) can be combined to provide

$$a_{11}\cos^2\phi_s + 2\,a_{12}\cos\phi_s\cdot\sin\phi_s + a_{22}\sin^2\phi_s = \frac{1}{2\cdot t_N(0,0)\cdot V_{\text{NMO}}^2(\phi_s)}.$$

From equation (9.21), we obtain

$$a_{11}^\circ\cos^2\beta_0\cos^2\phi_s + 2a_{12}^\circ\cdot\cos\beta_0\cdot\cos\phi_s\cdot\sin\phi_s + a_{22}^\circ\sin^2\phi_s = [2\cdot t_N(0,0)\cdot V_{\text{NMO}}^2(\phi_s)]^{-1}, \tag{9.24a}$$

where every quantity is known from the seismic survey with the exception of a_{11}°, a_{12}°, and a_{22}°.

If we look upon these four equations as an overdetermined system of linear equations for a°_{11}, a°_{12}, a°_{22}, the determinant of the 4×4 system consisting of the four rows of coefficients a°_{11}, a°_{12}, a°_{22} and the constant term must be zero. This determinant contains v_N as the only unknown. So we can eliminate a°_{11}, a°_{12}, a°_{22} in a rather simple way and compute v_N in a straightforward manner.

9.2 Interval velocities from SAM velocities

The recursive algorithms for computation of interval velocities from either NMO velocities (V_{NMO}) or SAM velocities (V_{SAM}) are largely identical. The diffraction time surface of Fig. 7-3 can be described in terms of the hypothetical D-wavefront that originates at D. The D-wave is, for the computation of interval velocities from SAM velocities, the counterpart of the NIP wavefront used for the computation of interval velocities from NMO velocities.

The recovery of interval velocities from V_{SAM} can be viewed in terms of having the D-wavefront shrink back into its hypothetical source at D. Now, however, we follow the image ray rather than the normal ray. For reasons of simplicity, we exclude from the following discussion those diffraction time surfaces that have more than one apex.

Since reflecting velocity boundaries and interval velocities can be recovered from SAM velocities and two-way primary image times, one could, for instance, compute V_{NMO} indirectly from V_{SAM} or vice versa using the recovered velocity model. This conversion is frequently required in seismic processing.

The process of obtaining interval velocities from time-migrated data is *significantly more efficient* than that for obtaining them from unmigrated (CDP-stacked) data for the following reasons:

(1) All image rays are traced vertically downward from the surface of the earth. Time-migrated horizons are more interpretable and often better resolved than CDP-stacked horizons; the picked image time of a reflector typically is, therefore, a more reliable quantity than the normal time. Also, velocity analyses based on time-migrated data are not confused with diffractions (*over-migrated* multiples, however, may still confuse velocity analyses).

(2) Image rays from one point on the earth's surface to different reflectors coincide with each other. They differ only in their lower end points. One can thus use all parameters computed along one image ray in the upper velocity layers (Δt_i, α_i, β_i, δ_i, $\mathbf{B}_i$) to solve for a deeper layer.

The recursive traveltime inversion algorithm can thus in fact be programmed without retracing new rays from the earth's surface to solve for deeper interfaces as is necessary for normal rays. The computation, therefore, is much simpler than that for normal rays. This is especially true for the matrices $\mathbf{B}_i$, which can be obtained particularly easily from second-order approximations of image time maps as discussed in section 8.2.

In working with normal incidence rays, one must obtain the interface shape over a range of CDPs before proceeding to the next interface down. No such problem exists with image rays. The algorithm can, in fact, be programmed without requiring any computer ray tracing through a model at all.

It is well known that V_{NMO} is generally close in value to the stacking velocity so long as velocity interfaces are fairly flat and the CDP spread length is not too long. The same relationship generally exists between V_{SAM} and migration velocities obtained for apertures that are not too large. When layer velocity boundaries are curved, however, it is quite possible that diffraction time surfaces no longer are sufficiently well approximated by hyperboloids, especially if the curvature of intermediate interfaces changes within the aperture range. In such a situation, however, because it is based on summing signals along hyperboloids, time migration *itself* will likely be unsatisfactory.

The problem of computing interval velocities from CDP traveltime measurements can thus be attacked with either of two complementary approaches, one using V_{NMO}, the other using V_{SAM}. Both methods exploit recursive wavefront curvature algorithms and are tailored to present-day recording and processing techniques. They are complementary in the sense that the V_{NMO} method will work best for simple reflections (i.e., relatively simple configurations of reflecting interfaces) and the V_{SAM} method in the case where diffractions dominate. Note that the V_{SAM} method cannot work at all in the case of plane layers when applied to CDP-stacked data.

One can alternatively look upon the two methods as providing redundancy to reduce uncertainties in the computation of interval velocities. Though, to date, little practical effort has been put into computing interval velocities from migration velocities, it is clear that the problems one will encounter should be similar to those found when computing interval velocities from stacking velocities.

9.2.1 Summary

NMO and SAM velocities can be related to the curvature of hypothetical emerging NIP wavefronts and *D*-wavefronts, respectively. The traveltime inversion methods are based on shrinking these hypothetical wavefronts back along either a normal ray or an image ray into their respective hypothetical sources. The inversion is obtained by means of a reverse recursion of the wavefront curvature laws.

The computation of interval velocities can thus be viewed essentially as a downward continuation process. The process is much like a TDM of a normal time map wherein a hypothetical normal wavefront shrinks back into a specific depth horizon.

When using NMO velocities, one recovers (for all available CDPs) first v_1 and the position of the uppermost reflecting horizon, then v_2 and the second horizon, etc.

On a regional scale, computed interval velocities will generally vary laterally, and the reflecting horizons will be obtained in the form of discrete points. Normal rays to deeper reflectors will intersect shallower velocity interfaces at interpolated layer boundary points which will have to be newly computed for each reflector.

On the other hand, when determining interval velocities from SAM velocities, one can solve for all layers at a chosen surface location at one time because image rays to successively deeper reflectors coincide with one another; furthermore, all desired interface curvature matrices can be obtained from second-order derivatives of a set of two-way image time maps at one and the same surface location.

9.3 Special velocity problems

In section 9.1, we showed that the interval velocity v_N becomes overdetermined when the NMO velocity V_{NMO} is known in more than one direction at a CDP point and the normal incidence times $t_{0,N}$ and gradient $\nabla\, t_{0,N}$ are known as well. This overdetermination pertains to the model of N isotropic constant velocity layers separated from one another by interfaces having arbitrary shapes. Does the overdetermination persist for a more general model? Let us consider three generalizations of the model.

First, we shall allow layers to be anisotropic. Then we shall discuss models in which the local velocity varies linearly in the vertical or in some other direction; finally, we shall approach the problem of determining the depth of the $(N-1)$th interface, under the assumption that no satisfactory reflection can be observed for this interface. The following paragraphs dealing with these three topics are included for completeness; they are treated only summarily in order to elucidate the problem, point to the unknown parameters and their relationship to the available data.

9.3.1 Anisotropy

We assume that the three terms of the second-order expansion for the CDP reflection time of the Nth horizon are known. Recall that for this purpose the CDP reflection times must be investigated in three directions. First, the algorithm of chapter 4 that describes the transfer of wavefront curvature (i.e., the second-order terms of the NIP wavefront) from the surface of the earth down to the NIP has to be adapted to this more general model. This generalization is possible in principle, and will result in three equations instead of one for the anisotropic velocities in the Nth layer (Gassmann, 1965, p. 75).

But are these equations sufficient to fully solve the problem of anisotropic velocities? Certainly not, because three parameters cannot completely describe a wave surface in anisotropic media (Krey and Helbig, 1956; Thomas and Lucas, 1977; Levin, 1978, 1979).

Even in the simplest case, i.e., that of a rotational ellipsoid which is the wave surface for genuine shear waves in the case of transversely isotropic media, we need two parameters defining the direction of the axis of rotational symmetry and two velocities (one in direction of the rotational axis and one perpendicular to it). A total of four parameters is required. Consequently, information at a single CDP is insufficient to solve the problem, even if V_{NMO} is known in all directions.

The determination of anisotropic velocities from observed reflection or diffraction times at, say, a few selected CDP points on the planar surface of the earth only carries other problems with it and cannot be solved unambiguously (Helbig, 1979). Helbig (1979, personal communication) points in this respect—for a special case—to a general theorem that implies that a wave surface agreeing with all observations at the surface of the earth can be transformed into other valid wave surfaces by affine transformations which leave the surface of the earth unchanged.

We do not want to go into more details here on this topic, the more so since we learned that Helbig will probably write a monograph on anisotropy.

However, two simple one-layer cases may be mentioned subsequently. If the wave surface is a rotational ellipsoid with the axis perpendicular to the surface of the earth, as is the case with genuine shear waves, a velocity computation at the CDP with a vertical normal ray yields the velocity which is valid in the directions perpendicular to the axis when the computation is based on the same assumptions as those used for an isotropic medium.

With quasi-dilatational waves in a transverse isotropic medium, the result essentially differs in this special case. Here we get a velocity value which is much closer to that in the direction of symmetry; it is even precisely that value if the anisotropy originates from laminated layering where for each individual isotropic lamina Poisson's constant σ equals $1/4$, as is often, at least approximately, the case (Krey and Helbig, 1956; Thomas and Lucas, 1976; Levin, 1978, 1979). Let us now return to the case of isotropy.

9.3.2 Linear velocity variation

Consider first a linear increase of velocity with depth z (Figure 9-6). For the sake of simplicity, the one-layer case is assumed in the following. Suppose the reflecting base dips in the x-direction. This direction is well known from ∇t_0, the gradient of the two-way normal traveltime (so long as we are dealing with a single-layer problem). If the NMO velocity V_{NMO} is known in two independent directions, we can compute it for all other directions—in particular for the x- and y- directions. Let the NMO velocities in these directions be $V_{NMO,x}$ and $V_{NMO,y}$. Let β_0 be the angle of emergence of the normal ray at the surface (Figure 9-6) and let

$$v(z) = v_0 + g\, z \tag{9.25}$$

be the linear velocity law. Then, provided we know t_0, ∇t_0, $V_{NMO,x}$, and $V_{NMO,y}$, we can compute v_0 and g as well as the depth and the dip of the reflector.

First, we can use the familiar formula

$$\sin \beta_0 = \frac{1}{2}\, v_0\, (\partial t_0/\partial x). \tag{9.26}$$

Note that for the linear velocity law, rays are circular arcs, and wavefronts from a point source are spheres in that case. From the fact that the NIP wavefront is a sphere, the following equation can be derived (see Appendix F).

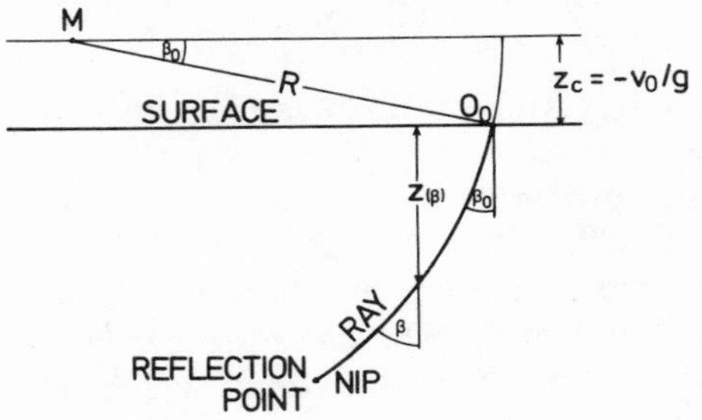

FIG. 9-6. Raypath geometry for a linear velocity increase with depth.

$$\cos\beta_0 \, V_{\mathrm{NMO},y}^{-2} - (\cos\beta_0)^{-1} \, V_{\mathrm{NMO},x}^{-2} = g t_0 \left(\frac{\partial t_0}{\partial x}\right)^2 \Big/ 8. \tag{9.27}$$

Equations (9.26) and (9.27) contain β_0, v_0, and g. Hence we can express the desired quantities v_0 and g in terms of β_0. We show in Appendix F that $V_{\mathrm{NMO},y}$ (the NMO velocity in the direction of strike) can easily be expressed analytically in terms of β_0, v_0, g, t_0, and $(\partial t_0/\partial x)$ using the Dix formula for a smoothly changing velocity medium

$$V_{\mathrm{NMO},y}^2 \, t_0 = \int_0^{t_0} v^2(t)\, dt. \tag{9.28}$$

Evaluation of this expression provides the key to the solution. Accounting for the fact that v_0 and g can be expressed in terms of β_0, we obtain the following equation with β_0 as the only unknown quantity

$$\begin{aligned} &2 \arctan\left(\tan(\beta_0/2) \exp\{4 (\partial t_0/\partial x)^{-2} \cdot \right.\\ &\left.[\cos\beta_0 \, V_{\mathrm{NMO},y}^{-2} - (\cos\beta_0)^{-1} V_{\mathrm{NMO},x}^{-2}]\}\right) \\ &= \arccos\left[(\cos\beta_0)^{-1} V_{\mathrm{NMO},y}^{-2} \, V_{\mathrm{NMO},x}^{-2}\right]. \end{aligned} \tag{9.29}$$

For more details, see Appendix F.

Because $\cos\beta_0 = 1 + \mathbf{O}(\beta_0^2)$ when β_0 is small, the determination of β_0 (and, hence, v_0 and g) becomes very poor. Thus, the method described here can yield satisfactory results only when dips are substantial.

The gradient of linear velocity increase can be changed from the z-direction to any other *known* direction without impairing the possibility of solving the problem. We need only carry out a rotation of the coordinate system so that the new z-axis points toward the direction of maximum velocity increase. There is, however, no possibility of determining this direction along with v_0 and g from the V_{NMO} values at just a single point on the surface of the earth.

Must we then conclude that v_0, g, and the position of the plane reflector cannot at all be recovered from the CDP reflection time curve in some direction? The answer is yes, so long as we are limited to using only the zero-order and second-order terms in the CDP reflection time curve. If, on the other hand, offsets are large enough to provide accurate estimates of higher-order terms, iterative modeling of CDP rays and traveltimes for various offsets can, in principle, provide a solution (see e.g., Gibson et al, 1979).

9.3.3 Nonreflecting interface

Now, we shall have a glance at the problem of inferring the shape of a nonreflecting interface given observations from the next deeper interface. Let us first consider the case of observations from interface $N = 2$ below a nonreflecting horizontal interface having unknown depth z_1 and unknown velocities v_1 and v_2 above and below. These three unknown quantities cannot be uniquely determined from knowledge of the NMO velocities (in all directions) of the second interface at a single point on the surface of the earth.

Because we are confining attention to a two-layer model in which the first interface is horizontal, observations of $\nabla t_{0,2}$ (the gradient of the normal time of the second reflecting interface) and only two independent values of V_{NMO} suffice for computation of V_{NMO} in any direction. Thus, additional knowledge is necessary, or assumptions have to be made. For instance, v_1 may be inferred from first arrivals, or v_2 may be known from geologic evidence (e.g., one might have knowledge that the second layer consists of salt).

The mathematics involved is similar to that required for the case of linear velocity increase (see Appendix F) and will not be derived here. Once v_1 and v_2 are known, the position at depth of the nonreflecting interface can be inferred. Just as in the case of linear velocity increase with depth, the second interface has to have a substantial dip in order to obtain reliable solutions.

This method can still be applied if the depth of the nonreflecting interface is slowly varying so that the interface can be looked upon as being planar with very small dip within the range of maximum offset. Thus, broad and gentle structural features of the nonreflecting interface can also be detected by applying the method at various points at the earth's surface.

In the more general case when the nonreflecting $(N - 1)$th interface is inclined and curved, the mathematics of Chapter 9.1 yields a differential equation of second order for the depth $z_{N-1}(x,y)$ of the $(N - 1)$th interface. Solution of this equation is possible (in principle), provided that v_{N-1} and v_N are known and that V_{NMO}, the NMO velocity of the Nth layer, has been determined in a constant direction throughout an x, y-survey (Fig. 2-1). This differential equation, with appropriate boundary values, can be used to determine the shapes of complicated nonreflecting or poorly reflecting velocity interfaces such as the bases of thrust sheets or the boundaries of intrusions and extrusions. Krey (1978) has applied this approach to problems involving hidden interfaces like those bounding salt and ore bodies as well as bodies originating from gravity slides.

10 Stacking-velocity analysis

In the preceding chapters we largely made use of ray-theoretical concepts and emphasized various algorithms for computing interval velocities from traveltime measurements. Now we will consider some more practical aspects of computing interval velocities from CDP reflection time measurements. We will outline some procedures that should be adhered to in order successfully to apply the generalized "Dix-type" formulas established above to measurements obtained in a real (geo-) physical world. Foremost, our traveltime inversion algorithms require that the real earth be closely approximated by the models considered above.

As in many other inversion algorithms, "idealized" surface measurements are required; such measurements are derived from "real" surface measurements only after applying certain corrections. Geophysicists who have previously computed interval velocities from either stacking- or migration velocities know that, typically, a host of often maddening problems is encountered when working with real seismic data. Some of these problems result from applying invalid corrections to the surface measurements.

The most critical parameter that must be obtained from CDP reflections is the NMO velocity V_{NMO}. At best, it equals the stacking velocity V_s for infinitesimal small offset only. A number of factors influence and bias V_{NMO} and V_s. These quantities are commonly computed in the digital computer by a procedure known as a *stacking-velocity analysis* (SVA).

A successful SVA requires adequately preprocessing CDP gathers and choosing some coherency measure for the computation of *velocity spectra*. Picking, validation, and smoothing schemes may follow after the spectra are computed. In addition to providing estimates of V_s, these schemes generally also make use of stack times and time dips to increase the overall reliability of all picked values. Though SVA has been considered primarily for determining the NMO corrections needed in CDP stacking, it is also a basic source of the observations required for computing interval velocities. In this application, SVA must be subjected to special high-resolution and accuracy requirements.

Since in the presence of long spreads and curved velocity boundaries stacking velocities generally do not equal NMO velocities, the removal of the *spread-length bias* must be given particular attention. Even when all necessary corrections are performed and interval velocities are obtained from accurately estimated observations in the surface seismic data, the total process of computing interval velocities might still not be considered complete. The interval velocities must, in all instances, be further subjected to some final statistical averaging and uncertainty estimation schemes in order to increase their ultimate value for interpretation. We do not want to discuss all these topics in great detail. We will only review some important results described in a number of publications, most of which are also relevant to the migration-velocity analysis (MVA) described in the next chapter.

Factors affecting observed velocity estimates.—From a theoretical point of view, the models used in this work have a reasonable degree of complexity. They are, however, still very simple as compared with actual complexities within the earth; for instance, we have ignored thin layering, vertical and horizontal velocity gradients, near-surface or distributed velocity anomalies, faulting, and anisotropy. By now, readers will have gained an appreciation for the importance of the parameters V_{NMO} and V_{SAM}, the primary ingredients in interval velocity computation. To obtain V_{NMO} with the necessary accuracy may require considerable effort; to some extent this effort can be automated. In order better to appreciate problems related to computing interval velocities from actual seismic data, let us review the most important factors that can affect their computation.

Among the numerous factors are those associated with *data acquisition;* examples include maximum and in-line offset, stacking multiplicity, and source and receiver characteristics. In a marine environment, the influence of source and streamer depth as well as streamer feathering must also be considered.

Some factors relate to specifics of *wave propagation* such as static shifts and change of wavelet character with offset and reflection time. Further complexity arises when primary events interfere with other neighboring primaries or with multiples and diffractions.

Data processing parameters also can have considerable influence. These include *parameters of the SVA program* such as muting, time gates, velocity sampling, and choice of coherence measure. Furthermore, computations reflect the interpreter's bias in the selection of horizons for analysis, always with observations inevitably contaminated by noise and measurement error. Record quality, with regard to either reflections or diffractions, can, of course, range from good to very poor. The signal-to-noise ratio generally degrades with increasing reflection time, making reflection time estimation less accurate just when greater accuracy is needed.

The higher frequencies required to delineate layers are the ones most attenuated by the earth. Their amplitudes also decrease most rapidly in transit through thin layering. The influence of the bandwidth of reflections on V_s has been investigated by Stone (1974). He shows that a decrease in bandwidth generally is accompanied by a decrease in the value of V_s.

Stacking velocity is thus a parameter which is quite sensitive to a large number of factors.

The role of the seismic interpreter.—Many geophysicists may probably find the theory and the programming effort for our algorithms demanding. Once programs are established and a feeling for the importance of input and output parameters as well as, for instance, of the significance of (bicubic) spline functions has been gained, these programs can be handled with much the same ease as simpler ones based on the Dix equations.

Interpreters need not be concerned with the complexity of the algorithms, nor need they require intuitive understanding of the dependence of V_{NMO} or

Table 10-1. Acceptable velocity errors (after Schneider, 1971).

	Acceptable error	
Use of velocity	RMS velocity	Interval velocity
NMO corrections for CDP stack as currently practiced	2–10 percent	—
Structural anomaly detection: 100-ft anomaly at 10,000-ft depth	0.5 percent	—
Gross lithologic identification: 1000-ft interval at 10,000-ft depth	0.7 percent	10 percent
Stratigraphic detailing: 400-ft interval at 10,000-ft depth	0.1 percent	3 percent

V_{SAM} on the layer boundaries or layer velocities. Such understanding is likely only when one restricts consideration to simple homogeneous plane-layer velocity models such as those discussed by Levin (1971).

Interpreters who want to avail themselves of the above inversion algorithms should, however, be constantly aware of the well defined limitations of models used and the exact meaning of the input parameters required for executing the computations. It is their task to restrict application of the sophisticated algorithms to use on good quality input data, to concentrate on reducing uncertainties in measurements and to detect and verify conditions which may violate the assumptions. They should exercise interpretational judgment and measure the results against general geologic likelihood.

Discriminating between desired and undesired information and recognizing conditions that oppose the basic assumptions for the inversion algorithms can involve a significant expenditure of an interpreter's time. However, a substantial amount of this work can be automated. Interactive processing can, for instance, be of considerable help at several critical phases of an analysis. Interpreters must balance the cost of expending such an effort against the cost of not doing so; the quality of a derived subsurface model primarily depends upon the interpreter's effort.

Accuracy considerations.—Considering stacking velocities for the purpose of computing meaningful interval velocities puts large resolution and accuracy demands on an SVA. These demands can be satisfied only if the CDP gathers used for analysis are properly preprocessed. This subject is reviewed briefly in section 10.1. The choice of an adequate *coherency measure* is of considerable importance for computing velocity spectra. Coherency measures are discussed in section 10.2.

Depending upon the ultimate use, velocity accuracy requirements can range widely in seismic processing and interpretation. Table 10-1 represents one assessment of the accuracy requirements for stacking velocities. The table, taken from a review paper (Schneider, 1971), was constructed under the assumption of a model consisting of horizontal isovelocity layers. Listed are different uses of velocity and corresponding acceptable errors in RMS velocity and, where applicable, estimated interval velocity.

The interval velocities are calculated from RMS velocities with Dix's equation. An accuracy of 2 to 10 percent for stacking velocities or RMS velocity has been considered sufficient in conventional processing. The more accurate end of the range, however, may be required for character detail studies of broadband amplitude-preserved data as performed in processing hydrocarbon zones (Larner, 1974).

If it were not for computing seismic interval velocities, the subject "stacking velocity analysis" would be almost a closed chapter in seismic exploration. Uncertainties in picked stacking velocities, however, can give rise to largely amplified uncertainties in computed interval velocities. Consequently, any research aimed at further improving the resolution of an SVA remains a challenging task.

The estimation of meaningful interval velocities for purposes of TDM or lithological studies imposes strict accuracy requirements on stacking velocities—often within less than a half percent. Indeed, a requirement of three percent accuracy in interval velocity over a 400-ft interval at a depth of 10,000 ft is unachievable; it requires an error less than one in a thousand for the RMS velocities. To aim for such high resolution would not only be unrealistic but also naïve since, for the many reasons outlined above, a chosen model could never completely characterize the real earth.

The larger the interval thickness, the smaller will be the error in computed interval velocity. However, a velocity estimate over a thick interval is less interesting than that over a thin one. To compute velocities for thin intervals with some confidence, one must increase the statistical reliability of the observed surface measurements. The various means by which this can be done are described later.

To estimate analytically the accuracy of computed interval velocities, one has to start by estimating the uncertainties (e.g., the standard deviation of the scatter) of the arrival times of primary CDP reflections (Bodoky and Szeidovitz, 1972). After that, one can analytically estimate the accuracy of the derived stacking velocities (Al-Chalabi, 1974) and finally the accuracy of computed interval velocities (Shugart, 1969; Kesmarky, 1976; Nakashima, 1977b). Though much can be learned from such error analyses, they are no substitute for care taken in recording, processing, and correcting seismic data. Also, the constraints provided by geologic plausibility are difficult to include in a rigorous mathematical analysis.

Normally, for a flat layered earth, an error in interpreted RMS velocity at time T is magnified by a factor of about $1.4T/\Delta T$ into an error in interval velocity for the time interval ΔT. We feel that by reason of continuity, this order of error will be very much the same for somewhat more complex models. However, this belief remains to be investigated in detail. From lessons learned through computer modeling, we also feel that the accuracy requirements for all other parameters, such as two-way time and normal time gradient, are less stringent than those put on the stacking velocity.

Whereas ray theory is best suited to solving the problems discussed in previous chapters, *communication theory and statistics*, considered in the light of the presently available computer technology, are essential to get the best results from high-resolution SVA.

10.1 Preprocessing CDP gathers

Certainly, optimum parameters must be selected for the various processes following the computation of velocity spectra. One should first ascertain, however, that the program that computes the spectra will be provided with *suitably selected* and *well-preprocessed* CDP gathers.

The selection of suitable CDP gathers away from faulted or otherwise disturbed zones is a matter of routine. Having CDP gathers properly preprocessed essentially implies that the primary CDP reflections should be properly aligned along CDP reflection time curves and enhanced against a background of multiples, diffractions, and random noise. This can be achieved by various wavelet processing, signal enhancement, and static-analysis methods. Their description could be the subject of an extensive treatment in itself, from which we will refrain. Nevertheless we feel that the role of these topics should be briefly reviewed to emphasize that successful computation of interval velocities cannot be seen outside the context of general procedures for seismic high-resolution signal processing.

10.1.1 Filtering

There has evolved a suite of digital-filtering techniques that can be tailored to particular needs resulting from properties of the acquisition and recording systems as well as from the geologic complexity of a survey area. One can select from among debubbling, dereverberation, deghosting, wavelet shaping, zero-phase, band-pass, and velocity filters to simplify reflection events on traces in a CDP gather. Multiples and reverberations are persistent problems, particularly on marine seismic data. Often they can be discriminated on the basis of their low stacking velocity relative to that of interfering primary events.

It is a common practice to sum a few successive CDP gathers in order to improve the signal-to-random noise ratio. This effort at reducing noise often involves a trade-off between improved accuracy and degraded lateral resolution in velocity estimates. Simple addition of CDP gathers is justified only if layers are horizontal and lateral velocity gradients are reasonably small. Even then, the summing of CDP gathers should be performed only after static corrections have been applied. Attenuation of random noise can also be achieved by increasing the CDP multiplicity over a given spread-length.

In the presence of dip and curvature of velocity boundaries, CDP gathers ought not simply be added together; rather, complicated procedures have to be designed to incorporate several neighboring gathers into a single SVA. This aspect will not be discussed here, but it is obvious that computations of V_{NMO} and V_s for assumed models can be helpful in this respect. For reasons of simplicity, let us assume in our further discussions that primary reflections can be well detected in single CDP gathers.

10.1.2 Static corrections

As the CDP spread moves across low-velocity, near-surface anomalies, time variations are introduced into the traveltimes of CDP reflections at different offsets. The variations are due primarily to lateral velocity inhomogeneities or depth undulations in a low-velocity, near-surface layer.

Primary CDP reflection energy must be correctly aligned along the desired hyperbola-like curves to utilize fully CDP gathers for velocity analysis. The alignment of all CDP reflections can often be achieved in the form of a constant time shift applied to each trace. The simple shifts are justified when travel paths for all reflectors and all offsets are sufficiently vertical near the surface. This assumption is generally acceptable for thin weathering layers of low velocity. Time shifts can then be assumed to be *surface-consistent*, so all traces for a common shot location receive the same *shot-static correction*, and traces for a common receiver location receive the same *receiver static correction*.

Typically, static corrections are performed in two steps—a first correction for *field statics* and a second one for *residual statics*. Field static corrections are deterministic quantities based on uphole times, short refraction lines, first arrivals, and the picking of predominant reflections. Residual static corrections are statistically derived quantities that can dramatically improve the coherence of seismic reflections. They are best obtained by automatic analysis of traveltime information (Taner et al, 1974; Saghy and Zelei, 1975; Wiggins et al, 1976).

Automatic static analysis and correction procedures have progressed to the point where numerous approaches work well when the near-surface layers affecting the statics are of low velocity and are sufficiently thin. Thick weathering fill and deep water-depth variations give rise to violations of the assumption of surface consistency. As with long-wavelength statics (near-surface caused time anomalies having low spatial wavenumber) which are not readily separable from subsurface structural variations, these particular static problems are far from being solved and remain vexing to processing personnel and interpreters alike.

Imperfect static corrections leave CDP reflection-time curves erratic; the resulting best hyperbolic fits then result in misleading stacking velocities and wrong interval velocities. The effects can be devastating, as has been repeatedly shown by Prescott and Scanlan (1971), Schneider (1971), Miller (1974), Pollet (1974), and Booker et al (1976). Near-surface anomalies are often the most disruptive factor in a velocity analysis. In fact, the large swings that too often exist in computed stacking velocities made their use in the computation of interval velocities suspect for a long time. Now that the nature of these variations is better understood, the method has re-emerged as an accurate and powerful tool.

Though undesired in most respects, near-surface time anomalies can provide useful supporting evidence of subtle changes in the stratigraphy within the overburden. Obvious errors in the values of V_s resulting from anomalies in the overburden can, in fact, help in upgrading the long-wavelength components of static corrections (Paturet, 1977). A deconvolution method based on this idea was discussed by Lucas et al (1975).

Let us provide one simple example (Figure 10-1) to support this short discussion on anomalous overburden problems. The medium is horizontally layered. Each

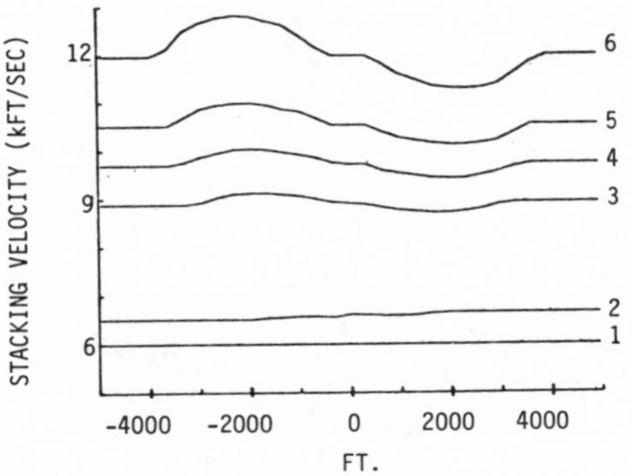

FIG. 10-1. CDP rays in a depth section including a low-velocity layer of abrupt change (after Larner, 1974).

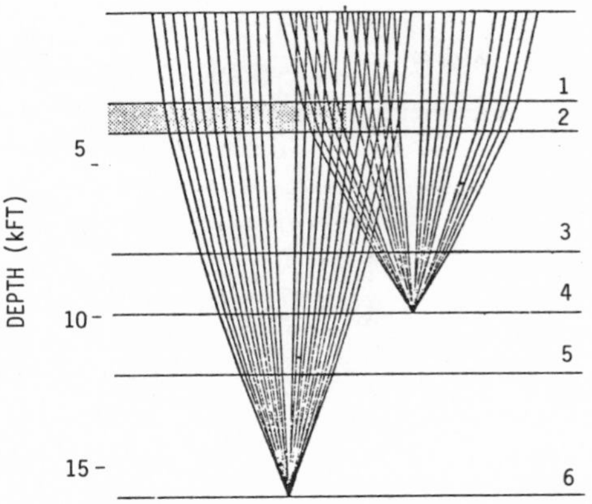

FIG. 10-2. Lateral variation of the stacking velocity obtained for the layer boundaries of the model of Figure 10-1 (after Larner, 1974).

velocity layer is homogeneous, with the exception of layer 2 in which the local velocity changes abruptly from 8000 ft/sec in the shaded portion on the left to 8500 ft/sec on the right. Raypaths are indicated for CDP locations on either side of the anomaly. For the CDP to the left, short-offset traces have raypaths that traverse only the low-velocity zone, whereas large-offset traces have raypaths that traverse the high-velocity region as well. CDP reflections observed on the large-offset traces thus arrive earlier than predicted from the small-spread hyperbola determined from the small-offset traces. Consequently, as shown in Figure 10-2, the stacking velocity (plotted as a function of CDP location) is unreasonably high for deep reflections to the left of the anomaly. Because stacking velocity becomes increasingly more sensitive to timing errors as depth increases, the velocity errors increase with increasing depth of reflectors. The errors extend laterally over distances approaching the maximum offset and have magnitudes that vary inversely with the square of it.

Although the anomalous stacking velocities may be proper for best alignment of primaries locally, any other use of them is not justifiable. For example, their use for converting from time to depth would result in large erroneous undulations of deep reflectors. Likewise, computed interval velocities would exhibit even more exaggerated variations.

The anomalous overburden problem provides the greatest rationale for a continuous determination of V_s along a seismic line. In the absence of continuous velocity coverage, anomalies in the overburden can often go unnoticed. For example, a stacking velocity function at an isolated analysis point located at −2000 ft in Figure 10-2 would be incorrect at 2000 ft or at 5000 ft. We would argue, in this monograph, in favor of continuous velocity coverage for the additional purpose of computing interval velocity in regions where curvatures and dips of interfaces vary continuously.

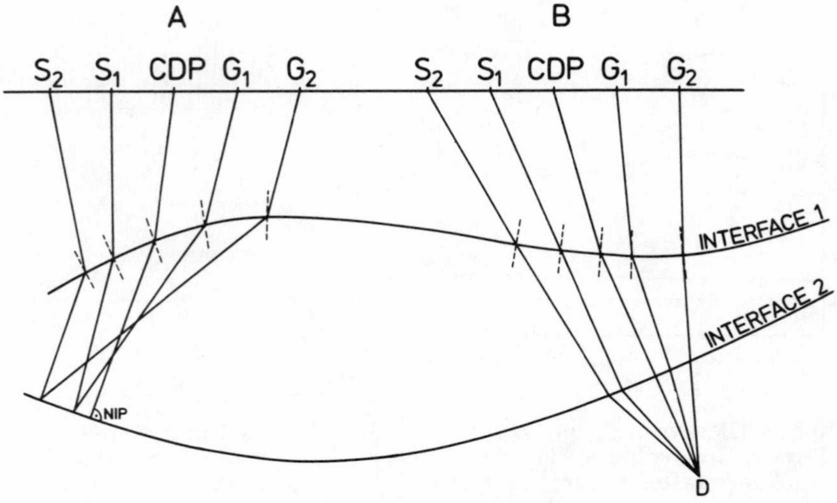

FIG. 10-3. CDP rays from two CDP profiles to a reflector and diffractor.

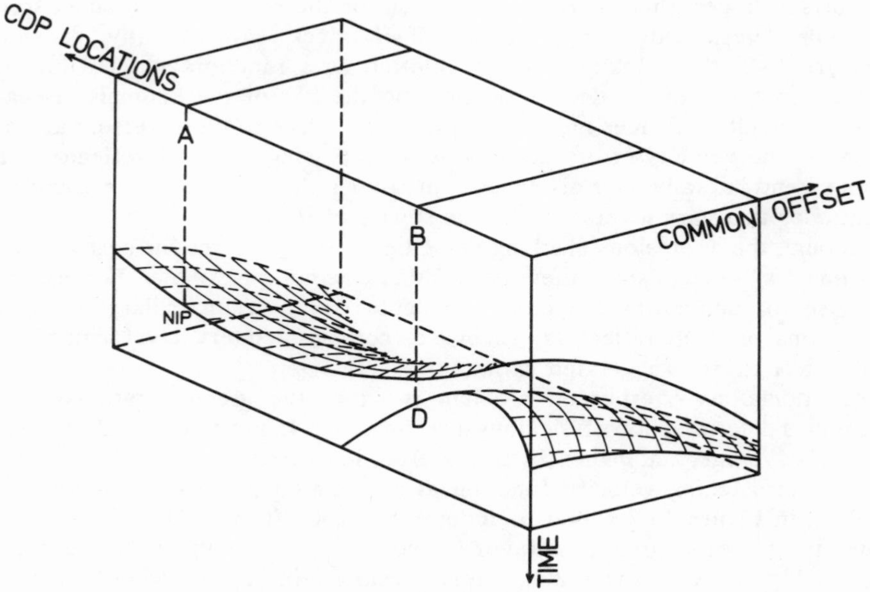

FIG. 10-4. CDP arrivals in CDP-time-offset space.

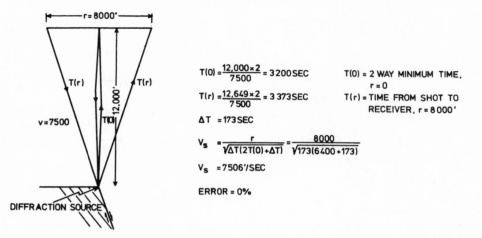

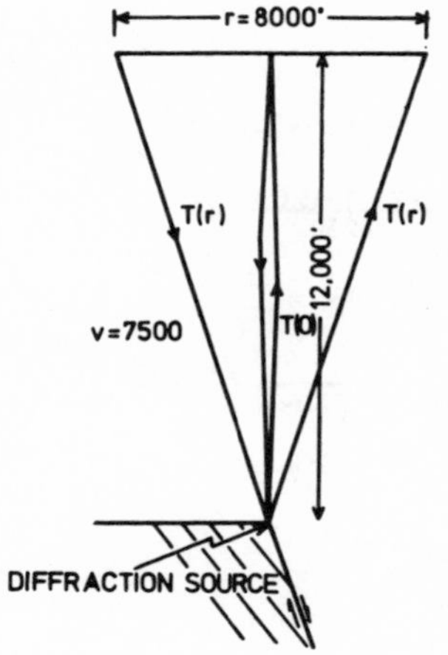

$$T(0) = \frac{12{,}000 \times 2}{7500} = 3200\,\text{SEC}$$

$$T(r) = \frac{12{,}649 \times 2}{7500} = 3\,373\,\text{SEC}$$

$$\Delta T = 173\,\text{SEC}$$

$$V_s = \frac{r}{\sqrt{\Delta T(2T(0) + \Delta T)}} = \frac{8000}{\sqrt{173(6400 + 173)}}$$

$$V_s = 7506'/\text{SEC}$$

ERROR = 0%

T(0) = 2 WAY MINIMUM TIME, r = 0
T(r) = TIME FROM SHOT TO RECEIVER, r = 8000'

FIG. 10-5. Velocity for a CDP spread centered over a diffractor (after Dinstel, 1971).

10.1.3 Diffraction patterns

Coherent hyperbola-like CDP arrivals need not all stem from energy specularly reflected at interfaces. Diffracting subsurface points give rise to similar hyperbola-like CDP arrivals. The diffraction curve is precisely hyperbolic, however, only when the medium is homogeneous and the CDP is directly above the diffracting source.

Often reflections and diffractions in a CDP gather can be discriminated from one another only with difficulty. Figure 10-3 shows CDP profiles at two different CDP locations *A* and *B* over a 2-D structure. Shown are the CDP ray families for the two kinds of energy return. On the left is a CDP ray family to the second interface and on the right is a family to a diffractor. Let us assume that the CDP spread moves from right to left. Figure 10-4 displays the CDP arrivals as surfaces in a space with axes representing offset, two-way time, and CDP location. One surface represents reflection times for interface 2 and the other diffraction times for the diffractor at *D*. Slicing the volume vertically parallel to the seismic line results in a *common-offset section*. Slicing it vertically parallel to the common-offset axis through a CDP point results in the *CDP record section* for that particular point.

For small offsets, primary events within a CDP record fall upon symmetric hyperbola-like curves regardless of whether they are associated with reflectors or diffractors. While diffraction amplitudes are generally weak, their implied velocity values can become anomalously large when the CDP spread moves away from the diffractor location.

The diffractor can be looked upon as a small reflecting sphere. As the CDP spread moves away from the sphere, the "reflection" comes from an increasingly steep portion of the "reflector surface." The increasing dip implies an increase in the (pseudo-) stacking velocity for which the diffraction event stacks best.

Figure 10-5 and Figure 10-6 feature such a situation for a diffractor below

$$T(0) = \frac{13.416 \times 2}{7500} = 3578 \, \text{SEC}$$

$$T(r) = \frac{12.168 + 15.626}{7500} = 3705 \, \text{SEC}$$

$$\Delta T = 127 \, \text{SEC}$$

$$V_{SP} = \frac{r}{\sqrt{\Delta T(2T(0) + T)}} = \frac{8000}{\sqrt{127(7156 + 127)}}$$

$$V_{SP} = 8319 \, '/\text{SEC}$$

$$\text{ERROR} = 11\%$$

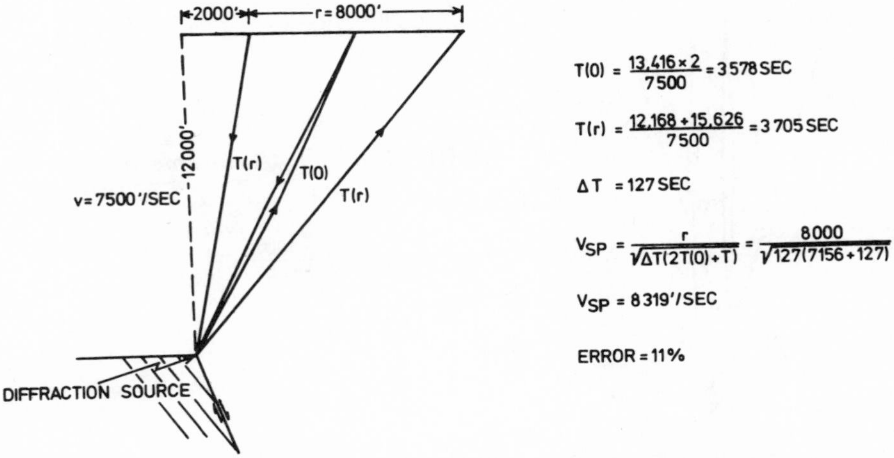

FIG. 10-6. (Pseudo) stacking velocity for a CDP spread not centered over a diffractor (after Dinstel, 1971).

a constant velocity medium. Diffraction stacking velocities can generally be discerned from stacking velocities if interpreters study them in conjunction with CDP-stacked and time-migrated sections (Dinstel, 1971). The lateral variation of time and velocity will reveal the lateral distance of a diffractor from considered SVA locations.

Stacking velocities for both diffractions and multiples constitute a similar kind of noise confusing the interpretation of true primary stacking velocity functions from velocity spectra. While stacking velocities for multiples are typically lower in value than those of primaries at comparable two-way times, pseudo stacking velocities resulting from diffractions are often higher.

The suppression of short-period multiples prior to CDP stacking is achieved partly by properly applied deconvolution methods. The suppression of diffractions prior to performing a stacking velocity analysis can be achieved by individually time migrating all common-offset sections (Doherty and Claerbout, 1976). However, the "analysis velocity" derived from such time-migrated common-offset sections leads to the stacking velocity as defined here only when the medium velocities do not vary laterally.

10.1.4 Summary

To give a good start to an SVA and the subsequent computation of interval velocities, one should properly select and preprocess the CDP gathers. In general, this implies performing some wavelet processing and applying *static corrections* when necessary.

Hyperbola-like CDP arrivals result from energy that is returned either by *reflection* or by *diffraction*. In a typical CDP gather, both kinds of energy return are included to a smaller or larger degree. Such energy can pertain to either multiples or primaries. While an SVA is most appropriate when energy is returned

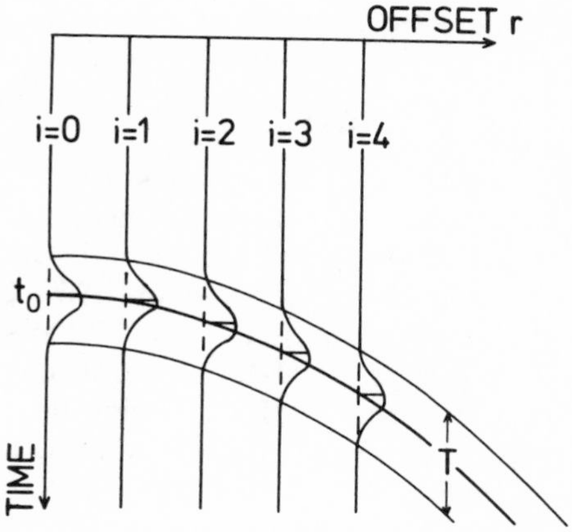

FIG. 10-7. Schematic reflection on a time-offset plane illustrating the principle of computing velocity spectra.

predominantly by specular reflection, *migration-velocity analysis* is more appropriate in the presence of subsurface bodies that backscatter seismic waves. In the presence of strongly curved reflectors, the distinction between reflected and diffracted energy fades.

10.2 Coherency measures

A good coherency measure for the detection of CDP reflections in preprocessed (NMO-corrected) CDP families is the human eye. Any newcomer concerned with SVA should first consider manual picking of CDP reflections before proceeding to make use of automated coherency measures.

Numerous techniques exist for routine extraction of V_s from CDP gathers. They are based on either summation of, or correlation between, CDP reflections and involve various choices of mathematical normalization. These techniques map coherent CDP reflection events from the time and offset domain into the zero-offset time and stacking-velocity domain. They replace the interpreter's eye for picking coherent reflections. The output map—*a velocity spectrum*—is typically a surface representing either reflection amplitude, correlation coefficient, or some other coherency measure versus zero-offset time and stacking velocity.

In this section we review, for completeness, some principles upon which the automatic computation of velocity spectra is based. The principles are made clear in Figure 10-7 which features a selected CDP reflection. At a particular *two-way time* t_0, a coherency measure is computed for various hyperbolic trajectories corresponding to a specified series of *test stacking velocities* $\bar{V}_s$ such that $V_{\min} \leq \bar{V}_s \leq V_{\max}$. A coherency value is computed for each test stacking velocity at the chosen time. The position of the maximum value of coherency then determines the optimum stacking velocity of a particular event.

By incrementing the t_0 time by an appropriate amount, we repeat a search such as this for different t_0 values over the entire record length of interest. From this effort, we can estimate the *optimum stacking velocity* as a function of the two-way zero-offset time.

The comparative analysis of coherence techniques for multichannel data is a broad topic. The most common coherency measures used for computing optimum stacking velocities have been subjected to numerous practical and controlled-performance tests (Garotta and Michon, 1967; Schneider and Backus, 1968; Taner and Koehler, 1969; Robinson, 1970b; Robinson and Aldrich, 1972; Montalbetti, 1973; and Sattlegger, 1975). Many of these measures are well summarized by Neidell and Taner (1971), in which the concept of *semblance* is introduced. Semblance, when properly interpreted, is a particularly powerful discrimination tool. We shall follow Neidell and Taner (1971) and Montalbetti (1973) in the discussion below.

Let us denote any digitized trace in a CDP-gather as $f_{i,t}$, where i refers to the channel index and t is the time index. The trajectory across the gather corresponding to a particular test stacking velocity $\bar{V}_s$ and two-way zero-offset time t_0 is denoted by $t\,(i)$, so that the stacked amplitude for M input channels is given by

$$s_t = \sum_{i=1}^{M} f_{i,t(i)} \,. \tag{10.1}$$

Since a trajectory need not pass through discrete sample points, $f_{i,t(i)}$ represents an interpolated value on the ith trace. The absolute value of s_t will exhibit a *maximum* when the trajectory $t(i)$ corresponds to the optimum stacking velocity V_s; that is, when the events are properly aligned before addition.

A normalized version of the expression (10.1) can be written in terms of absolute amplitude variation as

$$\text{COH} = \frac{\left|\sum_i f_{i,t(i)}\right|}{\sum_i |f_{i,t(i)}|} = \frac{|s_t|}{\sum_i |f_{i,t(i)}|} \,. \tag{10.2}$$

COH, the *normalized stacked amplitude*, takes on values ranging from unity when the signals are identical and have the same polarity, to zero when they are completely random or out of phase. Thus,

$$0 \le \text{COH} \le 1 \,. \tag{10.3}$$

Crosscorrelation is a related coherency measure. For each particular two-way time t_0 and velocity $\bar{V}_s$, zero-lag crosscorrelation sums are computed over moveout corrected windows of some specified length T centered about t_0. Since the actual trajectories of true reflections do not parallel one another, the window length T should be reasonably short. The crosscorrelation sum is determined according to

$$\text{CC} = \sum_t \sum_{k=1}^{M-1} \sum_{i=1}^{M-k} f_{i,t(i)} f_{i+k,t(i+k)}, \tag{10.4}$$

where the summations over k and i refer to all possible channel combinations and the sum over t refers to the time window over which the crosscorrelation is computed.

CC in equation (10.4) is an *unnormalized crosscorrelation sum* which can be written more simply as

$$\mathrm{CC} = \frac{1}{2} \sum_t \left[\left(\sum_i f_{i,t(i)} \right)^2 - \sum_i f_{i,t(i)}^2 \right], \tag{10.5}$$

or

$$\mathrm{CC} = \frac{1}{2} \sum_t \left(s_t^2 - \sum_i f_{i,t(i)}^2 \right).$$

Equation (10.5) shows the unnormalized crosscorrelation sum to be equal to half the difference between the output energy of the stack s_t and the input energy, where s_t is defined for trajectories (within the time gate) parallel to the one belonging to t_0. Thus, if $t(i)$ defines the trajectory of a coherent event across the input channels, the first term of equation (10.5) will be large with respect to the second, and CC represents a maximum.

Let us now consider normalized versions of equations (10.4) and (10.5). The usual normalization of equation (10.4) provides the *statistically normalized crosscorrelation sum*. This can be written as

$$\mathrm{CCN} = \frac{2}{M(M-1)} \cdot \sum_t \sum_{k=1}^{M-1} \sum_{i=1}^{M-k} \frac{f_{i,t(i)} f_{i+k,t(i+k)}}{\sqrt{\sum_t f_{i,t(i)}^2 \sum_t f_{i+k,t(i+k)}^2}}. \tag{10.6}$$

The factor preceding the sum ensures unit maximum amplitude, so that CCN varies between ± 1 according to the likeness and phase of the signals across the channels.

Expression (10.6) is a commonly used normalized crosscorrelation measure (the denominator is the geometric mean of energy in two channels over the time gate chosen). One can normalize expression (10.5) by the arithmetic mean trace energy instead of the geometric mean. If all the $f_{i,t}$ were equal, then equation (10.5) would reduce to

$$\mathrm{CC} = \frac{M(M-1)}{2} \sum_t f_{i,t}^2. \tag{10.7}$$

This behavior suggests definition of the *energy-normalized crosscorrelation sum,*

$$\mathrm{CCE} = \frac{\dfrac{1}{T} \dfrac{2}{M(M-1)} \mathrm{CC}}{\dfrac{1}{T} \dfrac{1}{M} \sum_t \sum_i f_{i,t}^2} = \frac{2}{M-1} \frac{\mathrm{CC}}{\sum_t \sum_i f_{i,t}^2}, \tag{10.8}$$

with

$$\frac{-1}{M-1} < \mathrm{CCE} \leq 1. \tag{10.9}$$

Computationally, expression (10.6) requires $M/2$ times more multiplications than does equation (10.8).

The statistically normalized crosscorrelation CCN will give unit correlation if the phases and shapes of the signal are identical across the channels, even if the CDP reflection wavelets in each trace are of different amplitude. Expression (10.8), however, will penalize such variations considerably. To understand this observation more clearly, consider the crosscorrelation for two channels $f_{1,t}$ and $f_{2,t}$ which have identical shape but differ in amplitude by a factor of a to 1 so that

$$f_{1,t} = f_t$$

and (10.10)

$$f_{2,t} = af_t.$$

The statistically normalized crosscorrelation is given by

$$\text{CCN} = \frac{\sum_t f_t \cdot af_t}{\sqrt{\sum_t f_t^2 \sum_t a^2 f_t^2}} = \frac{a \sum_t f_t^2}{a \sum_t f_t^2} = 1, \tag{10.11}$$

while the energy-normalized expression yields

$$\text{CCE} = \frac{\sum_t f_t \cdot af_t}{\frac{1}{2}\left[\sum_t f_t^2 + \sum_t a^2 f_t^2\right]} = \frac{2a}{1 + a^2}. \tag{10.12}$$

If, for example, $a = 0.5$ then CCN $= 1$ and CCE $= 0.8$ so that the sensitivity to amplitude difference is evident. (10.13)

Another coherency measure often used in SVA and MVA techniques relates to the ratio of stacked trace energy to input energy. Using equation (10.1), this ratio is defined as

$$\text{C} = \frac{\sum_t \left[\sum_i f_{i,t(i)}\right]^2}{\sum_t \sum_i f_{i,t(i)}^2} = \frac{\sum_t s_t^2}{\sum_t \sum_i f_{i,t(i)}^2}. \tag{10.14}$$

If there are M identical data channels, then

$$\text{C} = \frac{M^2 \sum_t f_t^2}{M \sum_t f_t^2} = M, \tag{10.15}$$

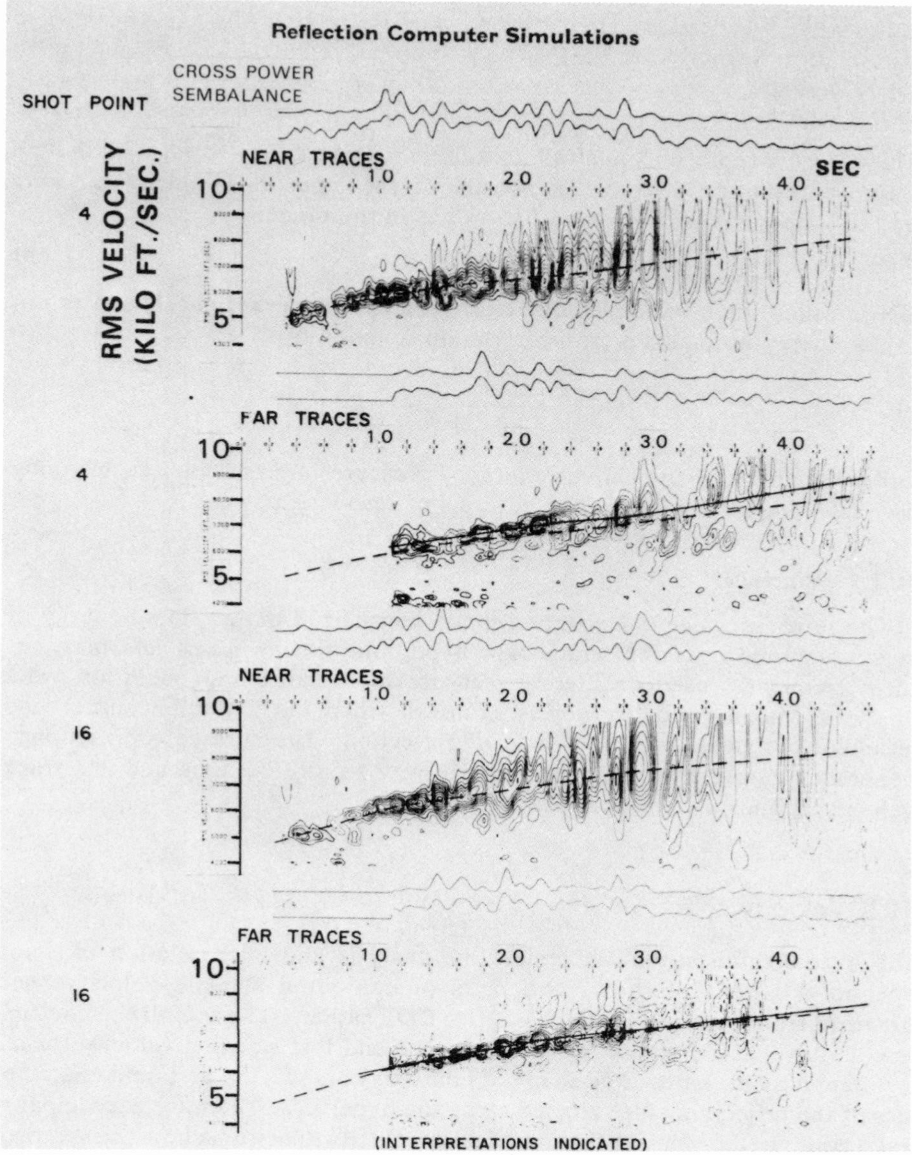

FIG. 10-8. Velocity spectra with interpreted stacking velocity functions superimposed (after Taner et al, 1970).

so that the normalized form of equation (10.14) becomes

$$\text{CN} = \frac{1}{M} \frac{\sum_t s_t^2}{\sum_t \sum_i f_{i,t}^2}. \tag{10.16}$$

This measure represents a *normalized output / input energy ratio,* with the output trace representing a simple composite of the input data which are centered on the trajectory $t(i)$. The ratio has values in the range

$$0 \le \text{CN} \le 1. \tag{10.17}$$

CN is called the *semblance coefficient* (Neidell and Taner, 1971) and is related to the energy-normalized crosscorrelation function by

$$\text{CCE} = \frac{1}{M-1}(M \cdot \text{CN} - 1). \tag{10.18}$$

Other variations of the above coherency measures are possible, each having its own advantages and limitations in terms of resolving power.

10.2.1 Summary

The *semblance coefficient* represents a normalized output/input energy ratio. It is computed over the width of a hyperbolic time window and may be the most frequently used coherency measure in stacking and migration velocity analysis. Maximum coherency is achieved when the hyperbolic time window straddles the peak energy of a CDP reflection. The surface representing the coherency measure as a function of two-way zero offset time and test stacking velocity defines a velocity spectrum.

10.3 Velocity spectrum

For completeness, let us review the analysis and interpretation of velocity spectra. Figure 10-8 shows two pairs of contoured stacking-velocity spectra obtained from near and far traces of two CDP gathers. The semblance coefficient is contoured as a function of two-way time and test stacking velocity for small overlapping time gates. The shape of contours contains the pertinent characteristics of the reflection pulse aligned along the hyperbola. The traces accompanying each contoured display at the top of each velocity spectrum show the semblance and crosspower maxima for each family of searches at the specified t_0 time.

The heavy solid line gives an interpretation (i.e., the *stacking-velocity function*) obtained from the far-trace spectrum. The dashed line gives the interpretation of the near-trace spectrum. (For comparison, both interpretations are displayed in the far-trace spectra.) Differences in the *stacking velocity* and *stack time* of events are clearly discernible. Stacking velocities for the far traces are as much as 500 ft/sec larger than those for the near traces.

The *spread-length bias* in this real-data example makes it clear that, when talking about stacking velocities, one should specify the spread-length and the offsets of traces used for analysis. In section 10.3.4., we shall discuss methods of reducing the effects of this spread-length bias. Apart from its usefulness for constructing a stacking-velocity function, the velocity spectrum is a powerful tool for discriminating between primary and multiple reflections.

In the absence of noise, semblance maxima coincide in time with the *peak energy* concentration of CDP reflections. Stacking velocities and stack times are thus necessarily *biased* by the *time delay* from reflection onset to peak energy concentration. The bias in both quantities is undesirable for the purpose of computing interval velocities since one wants information related to the *onset* of primary reflections. We shall refer to this type of bias as *onset bias.* Its neglect does not seem to be very critical for most time-to-depth conversions as the biases in the stacking velocity and stack time may largely cancel each other's effects (Everett, 1974). It should be mentioned that in Vibroseis traces after proper crosscorrelation with the sweep, the onset bias should be minimal.

10.3.1 Choice of SVA program parameters

To ensure stability in stacking-velocity estimates, time windows should overlap by about half their window length. They should not be shorter than about half the dominant period of the coherent wavelets of a CDP reflection. The window length will thus generally fall into the range from 20 to 80 msec depending upon the bandwidth of the signals. At the cost of loss in resolution, broadening the window length increases the eye appeal of a velocity spectrum by reducing the velocity fluctuations in peak positions and by reducing the confusion of weaker events. For detail studies, test stacking velocities are often incremented in steps of about 20 m/sec.

Muting early reflections (on the larger offset traces) is mandatory since wavelet character often changes strongly with increasing angle of reflection. Both offset dependence of wavelet character and possible interference with mode-converted and head waves can bias the stacking velocity unaccountably. Muting large-offset reflections generally does little harm to the determination of V_s for shallow reflections because the differential NMO times involved are large enough in the near traces to provide the necessary resolution in stacking velocity.

10.3.2 Picking and validation

Stack times and stacking velocities must be picked in a consistent manner. Because of their complexity, velocity spectra are most often interpreted manually. A stacking-velocity function is constructed by connecting those peaks in a spectrum which are considered to represent primary reflection energy buildups. To handle a large amount of data, the picking of coherency maxima is sometimes performed automatically by utilizing certain picking stratagems (Sherwood and Poe, 1972). One can, for instance, automatically pick all coherency maxima in a spectrum. This first step will provide both valid picks and spurious picks. The valid picks result from those with high probability of being primary events. The spurious ones result from picking minor lobes, multiples, diffractions, and aliases.

An automatic validation of picks can free seismic interpreters of the difficult decisions relating to the "goodness" of a pick. Picking and validation schemes generally incorporate not only stacking velocities but also stack times, normal time dips, and amplitudes. In this way, one quantity can be used to validate others relating to the same seismic event. Picking and validation schemes such as those discussed by Garotta (1971) and Donaldson (1972) are straightforward in concept. Nevertheless, much study and development is required before the digital computer will be able to interpret SVA as reliably as does the individual.

One problem, for example, occurs in areas where the stacking velocity of primaries does not increase monotonously with depth. Here multiples may have higher stacking velocities than do primaries at equal times. This occurrence is common, for instance, in the northern Alps and in permafrost provinces. It poses difficulties for human and machine alike.

To answer questions about *pick validity,* it is often essential to reexamine CDP reflections that produced a pick. Checking the consistency of amplitudes of a CDP reflection along the optimum stacking velocity trajectory can be helpful. When most amplitudes over the total offset range of a CDP reflection are comparable, a pick is more likely to be valid than when the amplitudes are less consistent.

Validation can be done in different stages. Once picks have passed various tests on the traces in a CDP gather, they can be further validated by subjecting them to continuity editing procedure involving stack times, normal-time dips, normal-reflection time curvatures, and amplitudes over a specified number of successive CDP records.

A way of further validating picked stacking velocities, time, time dips, etc., is to use all parameters for a "preliminary computation" of interval velocities. Only if computed interval velocities are "reasonable" are the picks that produced them retained. Interval velocities are thus not only the target of the basic interpretation effort, but also a means of better estimating the reliability of the picked events. Though some pick editing can be done automatically, it is strongly advised that interpreters monitor and control the process carefully, if possible by means of interactive graphics.

Automatically picked and validated stacking-velocity functions may finally be subjected to spatial smoothing. This process can reduce some statistical and systematic errors introduced by unrecognized or deliberately neglected minor irregularities in the overburden of the horizon in question. The effects of broad features in the velocity boundaries—broad as compared to the extent of a CDP profile—must, of course, not be removed by smoothing because they are taken into account in the inversion formulas that include dip and curvature of interfaces. It is essential that only picks belonging to a common horizon be used for smoothing. Smoothing computed interval velocities is conceptually simpler than smoothing stacking velocities because geologic judgment can better be used at that stage.

The picking, editing, and validation stratagems presently employed are designed primarily for seismic data collected within conventionally shot seismic lines. As outlined above, however, one must also know the crossdip (in time) of normal reflections to consider 3-D traveltime inversion algorithms (see section 9.1.3). Crossdip can be obtained from contoured zero-offset time maps. However, it may also be obtained as a field parameter from genuine areal or near-conventional strip profiling. Crossdip may have to be considered in future picking and validation schemes.

10.3.3 Onset bias

Let us assume a constant-velocity medium and a horizontal reflector for which a CDP reflection hyperbola has been obtained. If $\Delta\tau(r)$ describes the onset delay for the primary CDP reflection at offset r, and ΔV_s is the resulting error in V_s, then one can write

$$[t(r) + \Delta\tau(r)]^2 \approx [t(0) + \Delta\tau(0)]^2 + r^2/(V_s + \Delta V_s)^2. \qquad (10.19)$$

This is equivalent to

$$\Delta V_s / V_s \approx -[t(r)\Delta\tau(r) - t(0)\Delta\tau(0)] / [t^2(r) - t^2(0)]. \qquad (10.20)$$

Dix (1955) assumed $\Delta\tau(r)$ is constant, with the result that the error in V_s is minus one-half the proportional error in $t(0)$.

It may also be reasonable to expect the onset delay of wavelets in a CDP reflection to increase with offset since high frequencies are progressively attenuated with traveltime and wavelets thus increase in length. Assuming the onset delay to be proportional to traveltime, we get

$$\Delta\tau(r)/t(r) = \Delta\tau(0)/t(0). \qquad (10.21)$$

Together with equation (10.20), we obtain

$$\Delta V_s / V_s = -\Delta\tau(0)/t(0). \qquad (10.22)$$

This simple result confirms the intuitive feeling that an underestimate of velocity (resulting from a larger moveout) is counteracted by an overestimate of time. If the reflector depth is given by $d = V_s t(0)/2$, the proportional error in depth

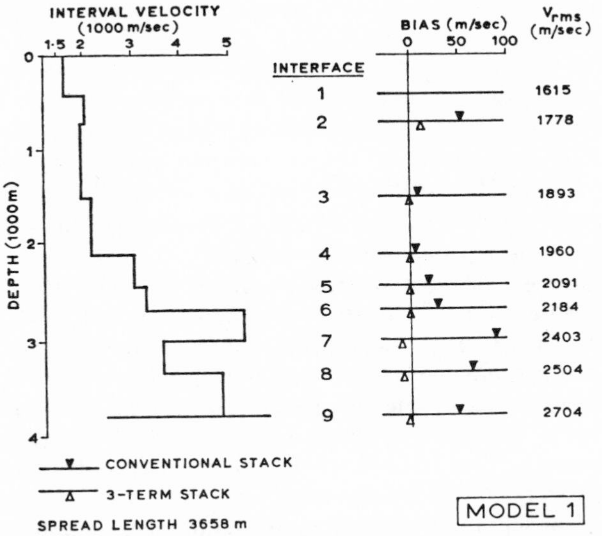

FIG. 10-9. Horizontal isovelocity-layer model (model 1) from the North Sea showing the spread-length bias in the estimate of V_{RMS} (after Al-Chalabi, 1974).

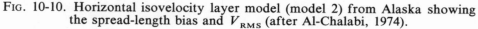

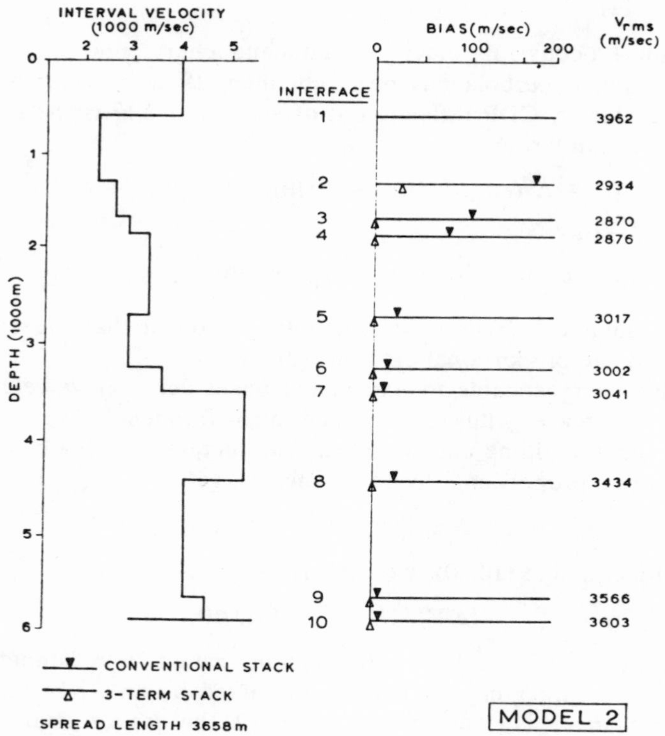

FIG. 10-10. Horizontal isovelocity layer model (model 2) from Alaska showing the spread-length bias and V_{RMS} (after Al-Chalabi, 1974).

d is given by

$$\Delta d/d = \Delta V_s/V_s + \Delta\tau(0)/t(0) = 0. \tag{10.25}$$

As the error in depth in this case is zero for the most simple model, it will likely also be small in more complex situations.

Though we may enjoy the favorable result of formula (10.25), we must not forget that the true onset is needed when computing interval velocities for correlation with a CVL log. Therefore, it may often be important to find an estimate for the onset correction $\Delta\tau(r)$ by visual inspection of the field records or by data processing. Note that on theoretical grounds onset problems should not be too severe in the Vibroseis-system as mentioned earlier.

10.3.4 Spread-length bias

The difference in values of NMO velocity and stacking velocity can increase with spread-length or, more precisely, with maximum offset. Unless allowed for, this difference can negate the advantages gained by the use of long spreads and cause large errors in interval velocity determinations. The spread-length bias can be reduced by a number of (largely automated) techniques (Brown, 1969; Al-Chalabi, 1973) some of which will be reviewed in the following.

Figures 10-9–10-10 provide two simplified horizontally layered velocity models from the North Sea and Alaska. Ray tracing has been used to simulate the exact onset times of primary CDP reflections from all velocity boundaries. Stacking velocities were simulated by least-square fitting hyperbolas through the actual CDP reflection time curves.

Both examples are taken from Al-Chalabi (1974). They have been selected to indicate the amount of spread-length bias one might expect. In general, the bias error is largest at and immediately beneath a zone of rapid increase in velocity. In model 1, the bias at a depth of 3000 m is much larger than at shallower horizons, say at 1500 m. This stresses the possibility of significant increases of bias with depth despite a decrease in the spread-length/depth ratio. In model 2, a large bias exists at shallow levels caused by the combination of large spread-length/depth ratio and sudden decrease in velocity below the permafrost layer. In this example, the bias decreases steadily with depth, except for an insignificant rise at the eighth interface.

When interval velocities are computed from stacking velocities that are assumed to be NMO velocities, the potential for large error is greatest where the spread-length bias changes significantly across an interval. For example, if no correction is made for the bias in calculating the interval velocity between the sixth and seventh interface of model 1 (Figure 10-9), the interval velocity will be overestimated by 14 percent. Likewise, the velocity of the interval between the fourth and fifth interface of model 2 (Figure 10-11) will be underestimated by 9 percent. These errors are too large for most purposes. They indicate the general need to take the spread-length bias into account when calculating interval velocities. The two methods described below are aimed at reducing this particular bias. They are based on procedures described by Al-Chalabi (1974) for the plane horizontally layered case, but can be applied to general 3-D cases as well.

The model simulation method.—It is assumed that all upper $N - 1$ velocity boundaries and layer velocities of a 3-D model (Figure 2-1) are known. For the Nth reflector only stacking velocities, stack times, and normal-time gradients are available at a number of CDP points. The bias between the desired NMO velocities and observed stacking velocities can be estimated by the following 3-step iterative procedure.

(1) V_s for each CDP location and arbitrary CDP profile is assumed to represent the wanted V_{NMO}. A first estimate for v_N and the depth of the Nth reflector is then worked out.

(2) Using the estimated v_N and estimated depth of the Nth reflector, a CDP ray family is then traced for each specified CDP profile. From the so-simulated CDP onset reflection times, a new stacking velocity $\bar{V}_s$ is computed at each CDP location by least-square fitting a hyperbola through the simulated CDP reflection time curve. An approximate value $\bar{V}_{\text{NMO}}$ is found by solving the forward traveltime problem along the normal ray for the approximate Nth reflector. An estimate of the actual bias $B = V_s - V_{\text{NMO}}$ is obtained from $B \approx \bar{B} = \bar{V}_s - \bar{V}_{\text{NMO}}$.

(3) $\bar{B}$ is then subtracted from the observed stacking velocity V_s, and $V_s - \bar{B}$ is considered to represent the desired V_{NMO} in step 1. Steps 1, 2, and 3 are then repeated several times until convergence to the correct values is achieved, and v_N and the Nth reflector are finally obtained.

Once stacking velocities and times are picked, the remainder of this iterative procedure can be fully automated. Hence, aside from the preliminary interpretation, interpreters need not be concerned with the difference between stacking velocities and NMO velocities.

Among alternative iterative schemes (Sattlegger, 1965; Davis, 1972; Gerritsma,

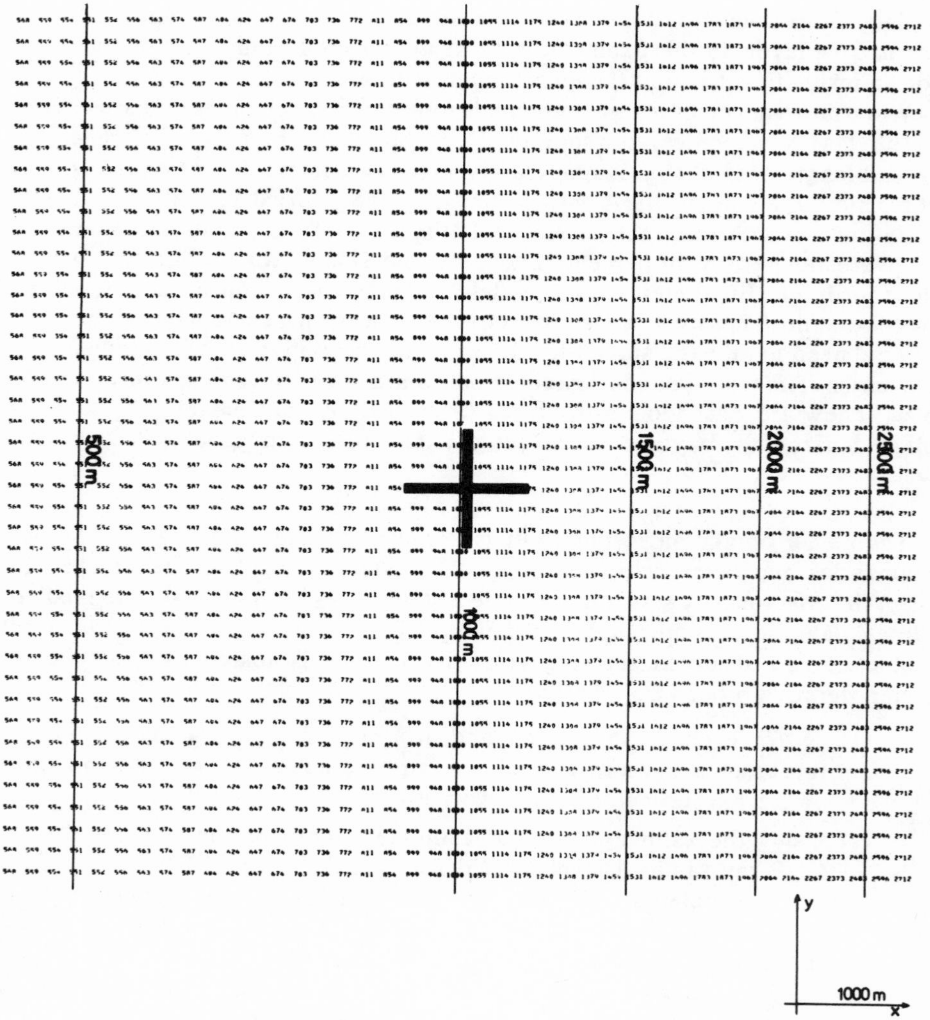

FIG. 10-11a. Contoured depth map for interface 1 of a 3-D model.

1977; Gjoystdal and Ursin, 1978), the following interpolation method is perhaps the simplest. Consider the following 3-D model of four constant-velocity layers studied by Meixner (1978). The contour levels (in meters) of the boundaries separating the layers are given in Figures 10-11a to 10-11d, and the CDP point is indicated by the cross in the middle of each plot. The layer velocities are $v_1 = 1600$ m/sec, $v_2 = 2400$ m/sec, $v_3 = 3600$ m/sec, and $v_4 = 5400$ m/sec, respectively. Figure 10-11e represents a plan view of the earth's surface into which eight CDP profiles of different azimuth directions are placed through the CDP.

The maximum offset r_{MAX} for all profiles is 7 km, and the subsurface coverage

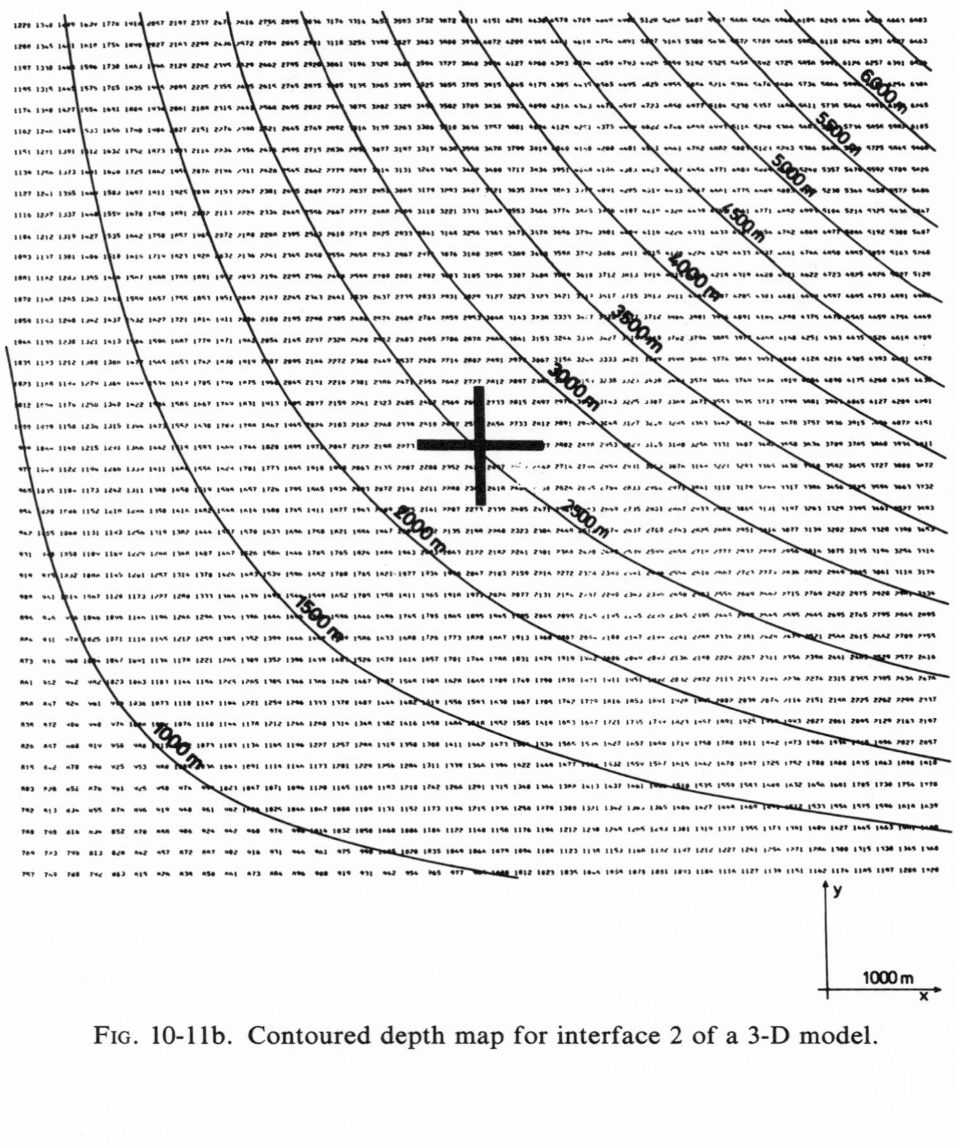

FIG. 10-11b. Contoured depth map for interface 2 of a 3-D model.

is now 36-fold. A coincident source-receiver pair is included at the CDP, and the source-receiver distance is incremented in steps of 200 m. The primary reflected ray between each source and its receiver (symmetrically placed with respect to the CDP) has been traced to the fourth horizon. From traveltimes computed for each ray, the CDP reflection time curves were computed within each CDP profile. (CDP reflection time curves for four profiles are shown in Figure 10-11f.)

The values of V_s for the fourth horizon as a function of all selected azimuths are obtained by least-square fitting a hyperbola to each time curve. They are marked as the largest in value of the two small numbers alongside the profiles

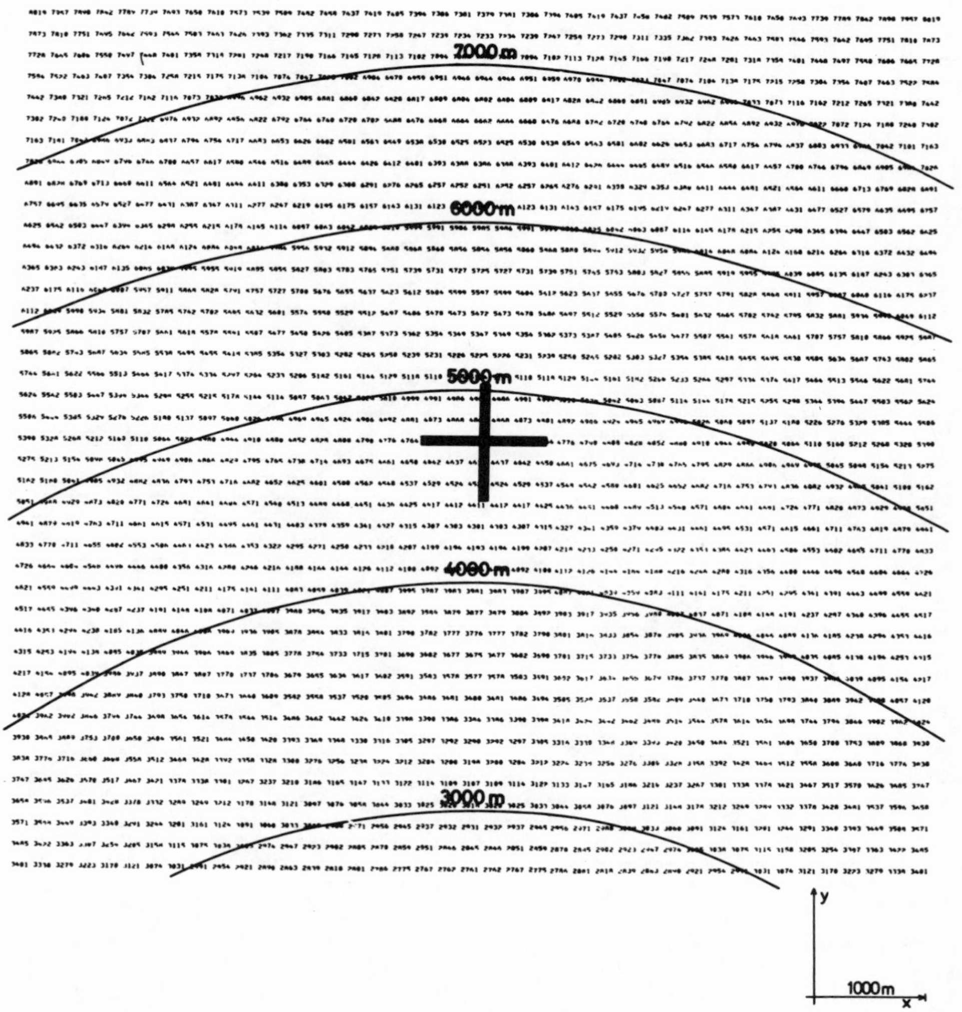

FIG. 10-11c. Contoured depth map for interface 3 of a 3-D model.

in Figure 10-11g. The stacking velocities to the third horizon are marked by the small number alongside the profiles. The mean deviation of the time difference between a hyperbola and its actual CDP reflection time curve was always less than 3 msec for the third and fourth horizon.

The large numbers in Figure 10-11g represent the interval velocities obtained from Dix's formula (when applied within each CDP profile) using the two-way normal times from the CDP to the third and fourth horizon (i.e., $t_{0,3} = 3.400$ sec and $t_{0,4} = 4.499$ sec). These computed interval velocities range with azimuth over more than 2000 m/sec.

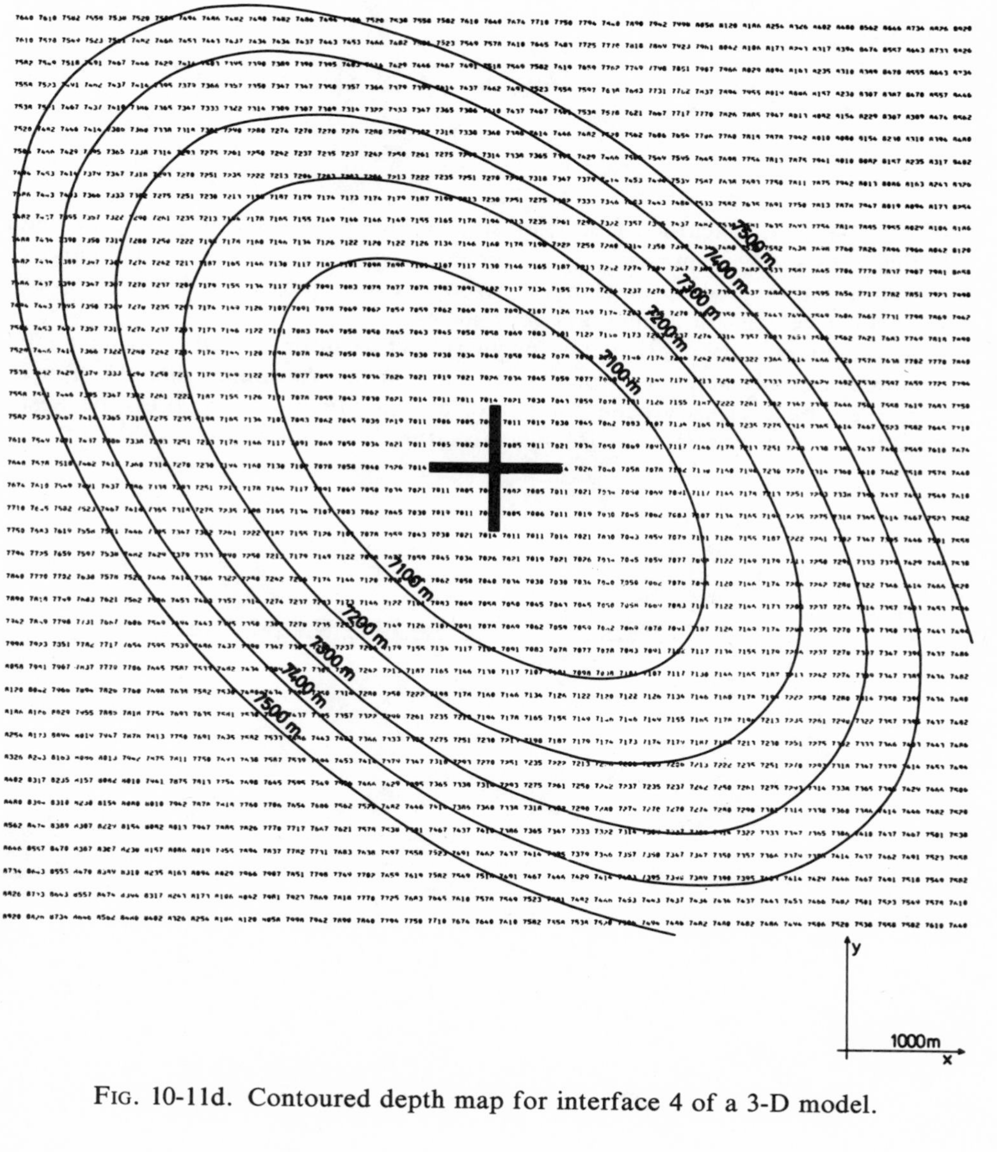

FIG. 10-11d. Contoured depth map for interface 4 of a 3-D model.

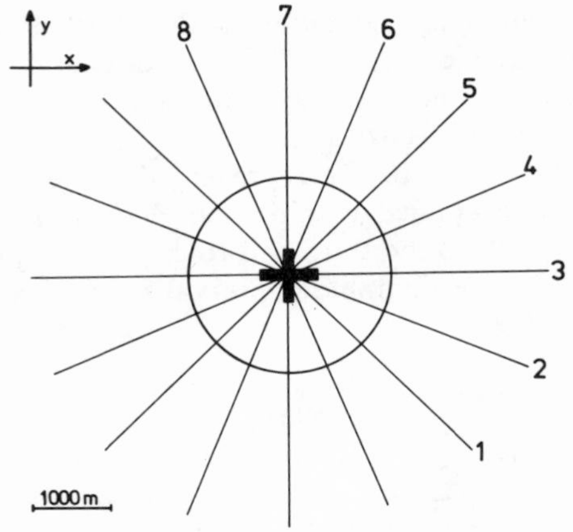

FIG. 10-11e. CDP profiles through the earth's surface.

What else can be learned from this simulation? It is interesting to note that for this reasonably complex velocity model, the CDP reflection time curves are well approximated by stacking hyperbolas even though the dependence of the stacking velocity on azimuth is very strong.

The scheme for recovering the fourth reflecting horizon and v_4 from CDP traveltime measurements was performed as follows. Assuming that the upper three layers have already been computed, an initial value for v_4 (i.e., the best available velocity estimate as, for instance, the Dix velocity unless this differs

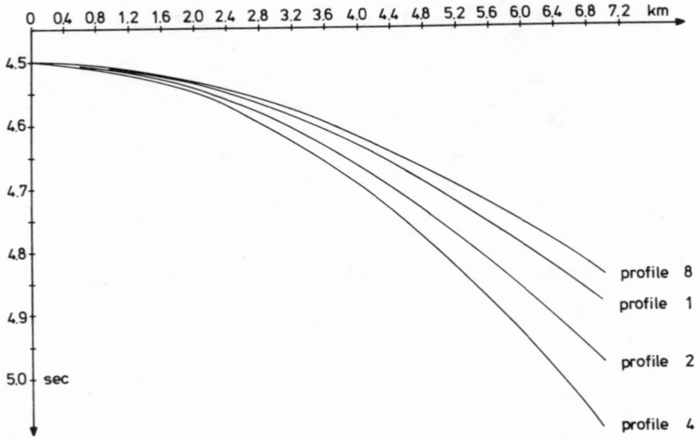

FIG. 10-11f. CDP reflection time curves of the fourth horizon for selected CDP profiles.

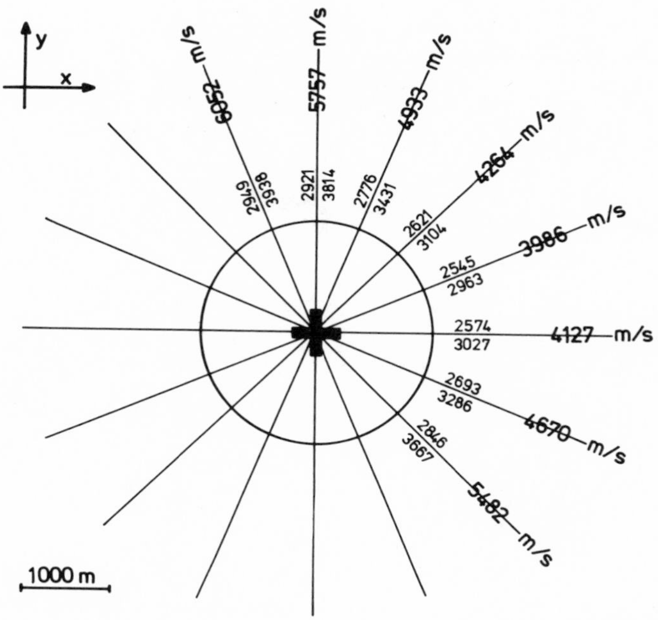

FIG. 10-11g. Stacking velocities to the third and fourth horizons (marked as small numbers close to circle) and Dix-computed interval velocities within profiles.

from a geologically reasonable value) is taken as the starting value for a sequence of iterations performed as follows.

Take the starting velocity v_4 and all available normal times of the fourth horizon to construct its approximate depth map by normal ray migration. Trace the CDP ray family to the approximate reflector within the specified profile and compute a first approximation of the observed CDP reflection time curve. Compare this computed time curve with the measured curve. If the computed NMO at r_{MAX} is larger than the actually measured NMO, the first assumed interval velocity is most likely too small. If the computed NMO is smaller than the actual NMO, then the first assumed interval velocity is most likely too large.

After computing NMO values at r_{MAX} for at least two different trial velocities v_4, one can obtain, by linear interpolation or extrapolation, a better value for the desired layer velocity. This value is then used for further iterative refinement. The iteration is stopped as soon as the computed NMO approximates the observed one within a specified time tolerance.

The iteration just described was applied within profile 1. It was started with the two velocities 4900 m/sec and 5900 m/sec. The interval velocity value 5412 m/sec was obtained after the fifth iteration when the difference in NMO at r_{MAX} was less than 1 msec. The difference in NMO remained less than that from then on. With 0.5 msec being the approximate accuracy of computed

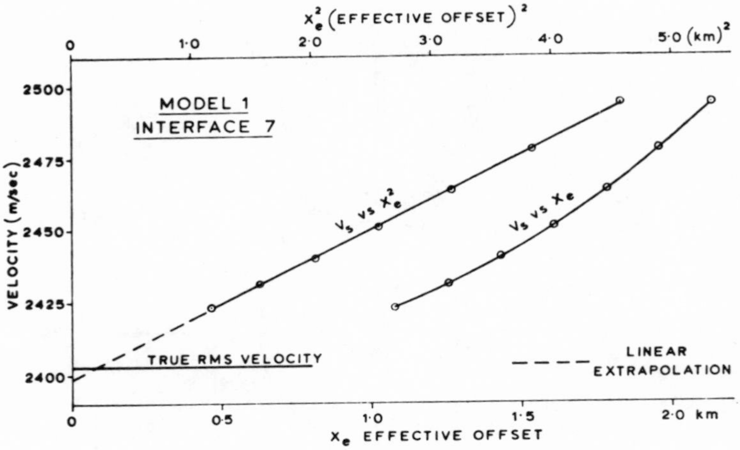

FIG. 10-12. Estimation of RMS velocity by a linear extrapolation from the V_s versus r_e^2 plot ($X_e = r_e$) (after Al-Chalabi, 1974).

time in the ray-tracing subprogram, the above-mentioned value of 5412 m/sec essentially cannot be improved by further iteration. Thus the final deviation from the true value of 5400 m/sec is about 0.2 percent (−54 dB).

The method of shifting stack.—The basic idea of this particular method is to estimate the NMO velocities from stacking velocities, in one series of steps, prior to solving for the velocity model. Basically, we extrapolate the value of V_{NMO} from a set of stacking velocities obtained for different CDP trace combinations at one CDP location. An important feature of this method in comparison to the above ones is that the estimate of V_{NMO} for a selected horizon is independent of the errors made in the velocity layers above.

Assuming 24-fold coverage, we can obtain a stacking velocity from analysis of the inner 12 traces (traces 1–12), then obtain another stacking velocity for the inner 14 traces (traces 1–14), another for the inner 16 traces, and so on until all 24 traces have been stacked. We would then end up with seven different stacking velocities for each CDP reflection.

The chosen traces could also be varied in a different manner; one could have stacked the traces 1–14, then 3–16, then 5–18, etc. This latter choice of combining traces is recommended in the presence of strongly curved interfaces where CDP reflections may not be well approximated by a hyperbola over the complete CDP profile length. Other combinations of offsets may be used. For instance, groups of traces with constant range of squared offsets may be the goal.

Let us define an *effective offset* as a single quantity that represents the shot-geophone offsets of all of the stacked traces. A convenient but arbitrary form for the effective offset is the RMS average offset

$$r_e = \left(\sum_{i=1}^{m} r_i^2 / m \right)^{1/2},$$

where r_i is the offset of the ith trace, and m is the number of CDP-stacked traces. We can plot the seven velocities versus their effective offset and extrapolate back to the zero offset velocity; i.e., NMO velocity. A low-order polynomial fitted to the points may be used in the extrapolation.

Figure 10-12 shows an example of V_s versus r_e^2 and V_s versus r_e plots over a horizontally layered ground. The V_s versus r_e^2 plot is recommended because theory suggests that it is almost a straight line. In practice, the plot will generally scatter (perhaps poorly) about a straight line mainly because of noise in the (reduced) stacked data.

Disadvantages of this method are (1) it requires considerably more computation of V_s estimates than does the iterative approach, and (2) each of the V_s functions has to be picked; moreover, the same event must be picked from all sets of trace combinations.

10.3.5 Summary

For the purpose of computing interval velocities, an SVA provides the most important link between theory and practice. It is the critical link because it provides the actual observed velocities for use in the ray-theoretical traveltime inversion algorithms.

Whether SVA is a weak link or a strong link depends largely upon the quality of seismic reflection data and on the performance of all preprocessing and initial correction procedures. These procedures, in conjunction with the computation of the velocity spectra, their picking, validation, and bias correction schemes, make up the overall analysis system. The entire analysis effort should be monitored and controlled by seismic interpreters.

The care required in extracting V_{NMO} estimates should properly match the sophistication of the traveltime inversion algorithms. Mistakes made in an SVA can seldom be corrected in a subsequent operation. Rather, these errors often result in greatly exaggerated errors in computed interval velocities because the latter are sensitive to errors occurring in the analysis.

11 Migration-velocity analysis

Conceptually, migration velocity is similar to stacking velocity. Whereas, stacking velocity best characterizes the normal moveout of reflection times across a CDP gather of traces, migration velocity best characterizes the moveout of diffraction times both across different CDP gathers and across traces within those gathers. Hence, migration velocity should be the appropriate parameter for migrating seismic data. Unlike stacking velocity, migration velocity is insensitive to the dip of the reflector to which it pertains and hence is single valued in space and is spatially more correct. For horizontally layered media, the two forms of velocity coincide.

As indicated in the last chapter, accuracy requirements when stacking velocities are considered solely for the purpose of CDP stacking are generally less stringent than those when the goal is to compute interval velocities. Judging from the analogies between CDP stacking and time migration, one can conclude that a similar relationship exists between the accuracy requirements when migration velocities are used for time migration and those for the computation of interval velocities. Since time migration is still a fairly new seismic process, until now most emphasis in the estimation of migration velocities has been put on obtaining well-migrated time sections (Sattlegger, 1975).

One can anticipate that progress in the use of migration velocities will parallel that of stacking velocities. Undoubtedly, in the future, migration velocities will be considered more often for the computation of interval velocities; the tremendous success of the time-migration method appears to be a good omen for this expectation.

Computation of interval velocities from migration velocities will require a quality-control and data-handling effort similar to that put into computing interval velocities from stacking velocities.

Here we will briefly describe some aspects that should be considered when using migration velocities for the purpose of computing interval velocities. From theoretical considerations made above and the many obvious similarities between CDP stacking and time migration, one can readily deduce how MVA should be performed.

In the subsequent considerations, we assume that all traces used in an MVA are properly preprocessed in ways described in the previous chapter so that we can immediately concentrate upon the computation of *migration-velocity spectra.*

From what has been said above, one will appreciate that the migration velocity of a diffracting event approximately describes an ellipse when plotted in polar coordinates as a function of the azimuth. Only if the subsurface is laterally homogeneous and if singularities are disregarded can data recorded by an areal recording configuration (Figure 7-3) be optimally time migrated using only a single (azimuth-independent) migration velocity. Only under these circumstances

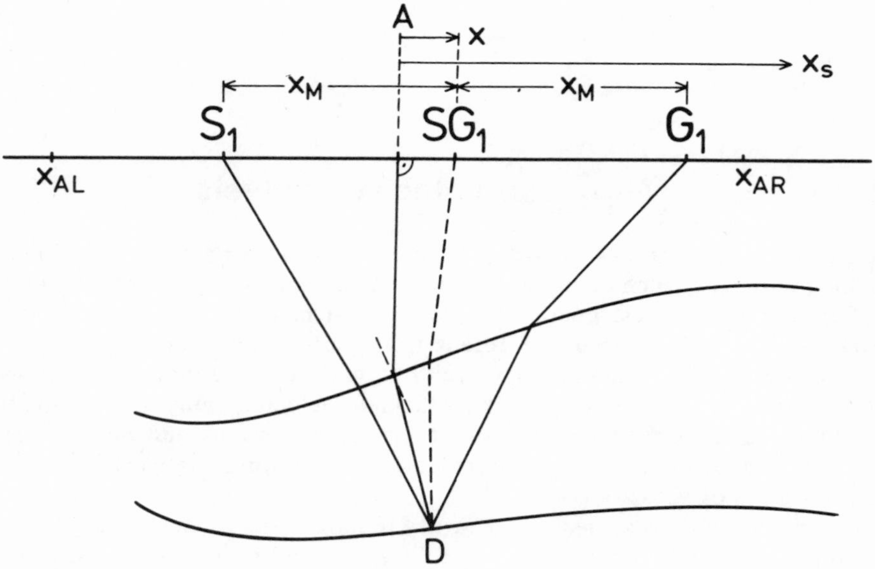

FIG. 11-1. Raypath for a member of a migration-before-stack gather used in 2-D migration velocity analysis.

can the diffraction time surface be approximated by a rotational rather than an elliptical hyperboloid.

For tutorial reasons, we first describe how MVA would be performed for a 2-D situation in which the profile is perpendicular to the common strike direction of all velocity boundaries. After that we outline the basic principles of MVA for a genuine 3-D earth model.

11.1 2-D Migration-velocity analysis

To obtain maximum reliability for the diffracted events in an MVA, we consider as many recorded, unstacked traces as are available within a specified aperture range.

Let the location of analysis along a seismic line be at point A, in the middle of the aperture range bounded by x_{AL} and x_{AR} on the seismic line (Figure 11-1). The source-receiver pairs included in the analysis belong to CDP profiles that translate along the seismic line. One source-receiver pair S_1-G_1 is indicated. The midpoint of the two points is denoted SG_1.

If at location A (with abscissa $x_s = 0$), a time-migrated trace is to be constructed from all available traces, then all the information it will include will stem from scattering subsurface points distributed along the image ray. The double traveltime of the hypothetical D-wave that originates at D can then be approximated over the chosen aperture range with respect to the x-coordinate by either

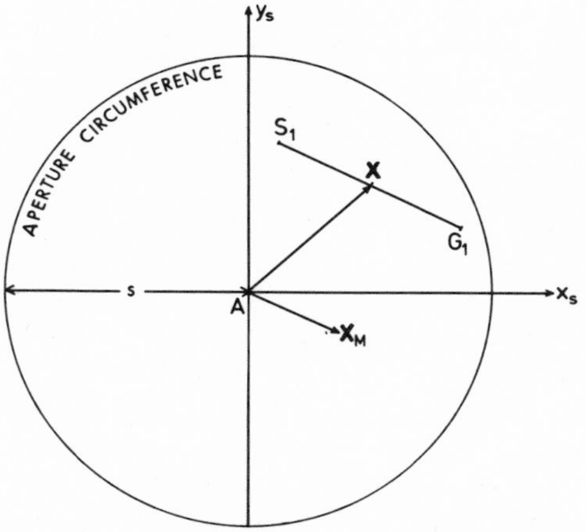

FIG. 11-2. Plan view showing the vectors **X** and $\mathbf{X}_M$ for a source and receiver within the aperture used for 3-D migration velocity analysis.

$$T_M(x) = T_M + 2x^2/T_M V_M^2,$$

or

$$T_M^2(x) = T_M^2 + 4x^2/V_M^2,$$

where V_M is the migration velocity and T_M is a time close to the two-way image time between A and D. T_M will be equal to the two-way image time if the aperture range is reduced to zero.

Let $T(x,x_M)$ be the scatter time from S_1 at abscissa $x - x_M$ to D and back to G_1 at abscissa $x + x_M$. Here, x is the abscissa of the midpoint between S_1 and G_1. $T(x,x_M)$ can then be approximated by

$$T^2(x,x_M) = T_M^2 + 4(x^2 + x_M^2)/V_M^2 = T_M^2 + 4x_S^2/V_M^2, \tag{11.1}$$

where $x_S^2 = x^2 + x_M^2$ and higher than second powers of x_S are neglected.

For $x \rightarrow 0$ and $x_M \rightarrow 0$, the migration velocity V_M approximates V_{SAM}. All traces in CDP gathers having midpoints in the range x_{AL} to x_{AR} constitute an augmented gather which we call a *migration-before-stack gather*.

This gather can be scanned for migration hyperbolas at any given two-way image time in the same way as a CDP gather is scanned for stacking velocity hyperbolas at two-way normal times. The migration velocity as a function of two-way time can then be extrapolated (as described above for V_s) to extract the small-aperture migration velocity V_{SAM}. The quantity can then be used for computing interval velocities by the methods described above.

The case of *migration after stack* is, of course, also included in the preceding considerations when $x_M = 0$. But in this case, no velocity information can be obtained in the plane layer case for pure specular reflectors where diffractions are absent. However, in the case of several different x_M-values, V_M can also be determined by equation (11.1) for plane purely reflecting beds (which may slightly dip so that the cosines of the normal ray emergence angles can still be assumed to be close to unity).

11.2 3-D Migration-velocity analysis

On the earth's surface (Figure 11-2) consider a number of source-receiver pairs at arbitrary locations placed such that their midpoints fall within a predetermined circular aperture range around the specified 3-D MVA location A. Again we desire to perform MVA on unstacked data and consider all source-receiver pairs within the specified aperture range. A time-migrated trace constructed at A displays information pertaining to points along the image ray through A. The diffraction time surface for the point scatterer D can be approximated by the following migration hyperboloid

$$T_M^2 (x_s, y_s) = T_M^2 + \frac{2\,T_M}{v_1} \mathbf{X}_s \bar{\mathbf{A}}_0 \mathbf{X}_s^T, \tag{11.2}$$

where $\mathbf{X}_s = (x_s, y_s)$ and

$$\bar{\mathbf{A}}_0 = \begin{bmatrix} \bar{a}_{11} & \bar{a}_{12} \\ \bar{a}_{12} & \bar{a}_{22} \end{bmatrix}.$$

T_M is approximately the two-way image time and v_1 is the velocity of the first layer. Higher than second powers of $|\mathbf{X}_s|$ are again neglected.

Equation (11.2) is not to be confused with equation (7.3) which describes the *SAM hyperboloid*. Equation (11.2) reduces to the SAM hyperboloid as the aperture reduces to zero; i.e., as $\bar{\mathbf{A}}_0$ approximates $\mathbf{A}_0$, the curvature matrix of the emerging D-wavefront at point A. The migration velocity can be expressed as

$$1/V_M^2 (\phi) = (T_M/2v_1)\, \mathbf{e}\, \bar{\mathbf{A}}_0\, \mathbf{e}^T, \tag{11.3}$$

with

$$\mathbf{e} = (\cos \phi, \sin \phi),$$

where ϕ describes some profile azimuth through analysis point A with respect to the $[x_s, y_s]$ system.

One source-receiver pair S_1-G_1 considered for a 3-D MVA at A is shown in Figure 11-2. The midpoint of the pair is at $\mathbf{X} = (x, y)$. The source S_1 is located at $\mathbf{X} - \mathbf{X}_M = (x - x_M, y - y_M)$ and the receiver G_1 at $\mathbf{X} + \mathbf{X}_M = (x + x_M, y + y_M)$. Let $T\,(x, x_M, y, y_M)$ be the scatter time from S_1 to D to G_1. It can be approximated by

$$T^2 (x, x_M, y, y_M) = T_M^2 + \frac{2\,T_M}{v_1} (\mathbf{X}\, \bar{\mathbf{A}}_0\, \mathbf{X}^T + \mathbf{X}_M\, \bar{\mathbf{A}}_0\, \mathbf{X}_M^T), \tag{11.4}$$

neglecting higher than second powers of $|\mathbf{X}|$ or $|\mathbf{X}_M|$.

This expression for T is a function of four scalar unknowns, T_M and the

Table 11-1.

SVA	MVA
Stacking velocity (V_s)	Migration velocity (V_M)
Normal moveout velocity (V_{NMO})	Small aperture migration velocity (V_{SAM})
Maximum offset (r_{MAX})	Half-width aperture (a)
Stacking hyperbola	Migration hyperbola
Normal ray	Image ray
CDP gather	Migration-before-stack gather

three elements of $\bar{\mathbf{A}}_0$, that need to be detected in order adequately to measure the migration velocity $V_M(\phi)$ for time T_M. Equation (11.4) is the 3-D counterpart of equation (11.1). When we reduce the aperture range, the migration velocity $V_M(\phi)$ we obtain from equations (11.4) and (11.3) reduces to the desired small-aperture migration velocity $V_{\mathrm{SAM}}(\phi)$.

In the most general case, the seismic field survey provides an areal distribution of $\mathbf{X}$ vectors as well as $\mathbf{X}_M$ vectors. Then, T_M and the three elements of $\bar{\mathbf{A}}_0$ can be determined by a best fit of actual observations to equation (11.4). In doing this, note again that we have an overdetermination of v_N as in the case of areal CDP coverage. Finding the migration hyperboloid (11.2) of a diffraction time surface, however, is no simple task because three parameters must be determined for each T_M value of an event.

It is much simpler to attempt fitting a rotational migration hyperboloid to a diffraction time surface (Sattlegger, 1975; Dohr and Stiller, 1975). It must be mentioned, however, that this provides acceptable results only when the medium is sufficiently homogeneous to make depth migration unnecessary; that is, image rays are nearly vertical, and time migration is adequate.

If CDP-stacked traces only are considered (i.e., if $\mathbf{X}_M = 0$), in principle the elements of $\bar{\mathbf{A}}_0$ may still be determined from equation (11.4). But in this case, plane reflecting horizons do not contribute to the desired information because such horizons can be time migrated with equal accuracy using any velocity. Thus $\mathbf{X}_M = 0$ will give satisfactory results only in faulted or folded areas where diffractions are evident in the data. In general, MVA using only CDP-stacked traces does not provide a sufficiently sensitive measure of migration velocity.

In many areal seismic surveys, all $\mathbf{X}_M$ vectors have nearly the same orientation given by the azimuth angle ϕ, while an areal distribution of $\mathbf{X}$ vectors is available. In this case, the determination of $\bar{\mathbf{A}}_0$ is reliable only in the direction of ϕ. This result means that only $\mathbf{e}\bar{\mathbf{A}}_0\mathbf{e}^{\mathrm{T}}$, with $\mathbf{e} = (\cos\phi, \sin\phi)$, can be derived. This is, nevertheless, sufficient information to solve the inverse traveltime problems and in particular, to determine v_N, the velocity of the Nth layer (see the last paragraph of section 9.1.3).

The case of CDP-stacked data from a seismic line survey oblique to a known direction of strike of diffracting faults and folds was considered by French (1975) and by Houba and Krey (1976). The authors found that the optimum migration velocity for normal 2-D migration along the oblique direction has to be multiplied by $\sin\psi$, where ψ is the angle between the strike and line directions, in order to get the true medium velocity. This result is strictly valid only if the diffracting structural features do not plunge in their strike direction and the medium velocity is constant.

11.2.1 Summary

Migration-velocity analysis (MVA) is basically similar to stacking-velocity analysis (SVA). Each concept in an SVA has its counterpart in an MVA, the most important of which are summarized in Table 11-1.

Though MVA is based initially on the diffraction time surface which is defined for *coinciding* source-receiver pairs, we have shown how to derive the desired velocity and time parameters from a *migration-before-stack gather* involving separated sources and receivers. Often MVA before stack is a costly process that involves as many as 500 to 1000 traces. MVA after stack, however, has little value because most of the migration velocity information is lost in the conventional stacking process and because it fails completely in the purely reflecting plane layer case.

Appendix A
Transformations of coordinate systems and curvature matrices

Let $\hat{\mathbf{T}}(\alpha)$ be the following 3×3 matrix

$$\hat{\mathbf{T}}(\alpha) = \begin{bmatrix} \cos\alpha & 0 & \sin\alpha \\ 0 & 1 & 0 \\ -\sin\alpha & 0 & \cos\alpha \end{bmatrix}. \tag{A.1}$$

Here and in the following the symbol ˆ in connection with a boldface letter denotes a 3-D matrix (or vector). Otherwise, boldface letters denote 2-D quantities. Let us also introduce the following notation:

$$\hat{\mathbf{X}}_I = (x_I, y_I, z_I); \quad \hat{\bar{\mathbf{X}}} = (\bar{x}, \bar{y}, \bar{z});$$
$$\hat{\mathbf{X}}_T = (x_T, y_T, z_T); \quad \hat{\bar{\mathbf{X}}}_T = (\bar{x}_T, \bar{y}_T, \bar{z}_T),$$

and

$$\hat{\mathbf{X}}_R = (x_R, y_R, z_R); \quad \hat{\bar{\mathbf{X}}}_R = (\bar{x}_R, \bar{y}_R, \bar{z}_R);$$
$$\hat{\mathbf{X}}_F = (x_F, y_F, z_F); \quad \hat{\mathbf{X}} = (x, y, z),$$

where $\hat{\mathbf{X}}_I$ denotes the coordinates of a point within the $[x_I, y_I, z_I]$ system, $\hat{\bar{\mathbf{X}}}$ within the $[\bar{x}, \bar{y}, \bar{z}]$ system, etc. Coordinate transformations at points of reflection (refraction) can be expressed as follows

$$\hat{\mathbf{X}}_F = \hat{\mathbf{X}}_I \hat{\mathbf{T}}(\varepsilon_I), \tag{A.2}$$
$$\hat{\mathbf{X}}_F = \hat{\bar{\mathbf{X}}}_T \hat{\mathbf{T}}(\varepsilon_T), \tag{A.3}$$

and

$$\hat{\mathbf{X}}_F = \hat{\bar{\mathbf{X}}}_R \hat{\mathbf{T}}(\varepsilon_R - \pi). \tag{A.4}$$

Recall the sign convention used for ε_I, ε_T, and ε_R (see step 4 in section 4.2); ε_I and ε_T have equal sign, while ε_I and ε_R are of opposite sign.

The connection between $\hat{\mathbf{X}}_T$ and $\hat{\bar{\mathbf{X}}}_T$ or $\hat{\mathbf{X}}_R$ and $\hat{\bar{\mathbf{X}}}_R$ is described by the following transformations

$$\hat{\mathbf{X}}_R = \hat{\bar{\mathbf{X}}}_R \hat{\mathbf{D}}(\delta), \tag{A.5}$$

and

$$\hat{\mathbf{X}}_T = \hat{\bar{\mathbf{X}}}_T \hat{\mathbf{D}}(\delta), \tag{A.6}$$

where

$$\hat{\mathbf{D}}(\delta) = \begin{bmatrix} \cos\delta & -\sin\delta & 0 \\ \sin\delta & \cos\delta & 0 \\ 0 & 0 & 1 \end{bmatrix}. \tag{A.7}$$

Both $\hat{\mathbf{T}}(\alpha)$ and $\hat{\mathbf{D}}(\delta)$ are rotation matrices. A familiar property of rotation matrices is that the inverse of a rotation matrix equals its transpose.

Equation (4.21), which pertains to the $[\bar{x}, \bar{y}, \bar{z}]$ system at $\mathbf{O}_i$ can be written as

$$\hat{\hat{\mathbf{X}}} \hat{\bar{\mathbf{B}}} \hat{\hat{\mathbf{X}}}^T + \hat{\bar{\mathbf{C}}} \hat{\hat{\mathbf{X}}}^T + s_{i+1} = 0, \tag{A.8}$$

where

$$\hat{\bar{\mathbf{B}}} = \begin{bmatrix} \bar{b}_{11} & \bar{b}_{12} & 0 \\ \bar{b}_{12} & \bar{b}_{22} & 0 \\ 0 & 0 & 0 \end{bmatrix},$$

and

$$\hat{\bar{\mathbf{C}}} = (\bar{c}_1, \bar{c}_2, -1).$$

Let us now find the wavefront approximation (A.8) with respect to the $[x_F, y_F, z_F]$ system at $\mathbf{O}_{i+1}$. Using equation (A.5) or (A.6) and recalling that $[\bar{x}, \bar{y}, \bar{z}] \equiv [\bar{x}_R, \bar{y}_R, \bar{z}_R]$ or $[\bar{z}, \bar{y}, \bar{z}] \equiv [\bar{x}_T, \bar{y}_T, \bar{z}_T]$, whatever the case may be, we can write at $\mathbf{O}_i$

$$\hat{\hat{\mathbf{X}}} = \hat{\mathbf{X}} \hat{\mathbf{D}}^{-1} (\delta_i).$$

As the $[x_I, y_I, z_I]_{i+1}$ system at $\mathbf{O}_{i+1}$ is obtained from the $[x, y, z]_i$ system at $\mathbf{O}_i$ by a parallel shift of axes, we can rewrite this expression for $\hat{\hat{\mathbf{X}}}$, using (A.2) as

$$\hat{\hat{\mathbf{X}}} = \hat{\mathbf{X}}_F \hat{\mathbf{T}}^{-1} (\varepsilon_{I,i+1}) \hat{\mathbf{D}}^{-1} (\delta_i) + (0, 0, s_{i+1}).$$

By substituting this expression into equation (A.8), we finally can express equation (A.8) within the $[x_F, y_F, z_F]$ system at $\mathbf{O}_{i+1}$ as

$$\begin{aligned} &\hat{\mathbf{X}}_F [\hat{\mathbf{T}}^{-1} (\varepsilon_{I,i+1}) \hat{\mathbf{D}}^{-1} (\delta_i) \hat{\bar{\mathbf{B}}} \hat{\mathbf{D}} (\delta_i) \hat{\mathbf{T}} (\varepsilon_{I,i+1})] \hat{\mathbf{X}}_F^T \\ &+ \hat{\bar{\mathbf{C}}} \hat{\mathbf{D}} (\delta_i) \hat{\mathbf{T}} (\varepsilon_{I,i+1}) \hat{\mathbf{X}}_F^T = 0. \end{aligned} \tag{A.9}$$

Equation (A.9) must be equal to

$$\mathbf{X}_F \mathbf{B} \mathbf{X}_F^T - 2 z_F = 0. \tag{A.10}$$

By comparing coefficients in equation (A.9) and equation (A.10), one can express the desired 2×2 matrix $\mathbf{B}$ at $\mathbf{O}_{i+1}$ as

$$\mathbf{B} = 2 \mathbf{S}_I \mathbf{D}^{-1} \bar{\mathbf{B}} \mathbf{D} \mathbf{S}_I, \tag{A.11}$$

where

$$\mathbf{S}_I = \begin{bmatrix} \cos \varepsilon_{I,i+1} & 0 \\ 0 & 1 \end{bmatrix}; \quad \mathbf{D} = \begin{bmatrix} \cos \delta_i & -\sin \delta_i \\ \sin \delta_i & \cos \delta_i \end{bmatrix};$$

$$\bar{\mathbf{B}} = \begin{bmatrix} \bar{b}_{11} & \bar{b}_{12} \\ \bar{b}_{12} & \bar{b}_{22} \end{bmatrix},$$

and

$$\cos \varepsilon_{I,i+1} = 1/\sqrt{1 + \bar{c}_1^2 + \bar{c}_2^2} \quad ; \quad \tan \delta_i = \bar{c}_2 / \bar{c}_1.$$

Appendix B
Derivation of the refraction (reflection) law of wavefront curvature

Derivation of the curvature laws can be based on any one of various principles, the most appealing probably being the *phase-matching principle* used by Deschamps (1972). Our derivation here makes use of the concept of the "directional unit sphere" (Krey, 1976). It provides some insight into the laws not gained by other derivations. Let us start first with the *refraction law*.

Figure B-1 shows a refracted ray at an interface point $\mathbf{O}_T$. A second close ray, also shown, belongs to the same wavefront and pierces the interface at $\mathbf{O}_{T,\Delta}$. Along these rays, unit vectors are placed pointing in the direction of the advancing wavefront. $\mathbf{s}_I$ and $\mathbf{s}_T$ are the unit vectors along the incident and refracted ray through $\mathbf{O}_T$; $\mathbf{s}_{I,\Delta}$ and $\mathbf{s}_{T,\Delta}$ are the counterparts for the offset ray through $\mathbf{O}_{T,\Delta}$; v_I is the interval velocity at the incident and v_T at the refracted side of the interface; $\mathbf{n}_F$ is the interface normal vector at $\mathbf{O}_T$, and $\mathbf{n}_{F,\Delta}$ is its counterpart at $\mathbf{O}_{T,\Delta}$. As in the text, interface vectors point into the medium away from the arriving wave.

If all unit vectors are translated (without rotation) so that they originate at $\mathbf{O}_T$, their end points fall onto the directional sphere (Figure B-2). Snell's law confines the vectors $\mathbf{s}_I$, $\mathbf{s}_T$, $\mathbf{n}_F$ to one plane and $\mathbf{s}_{I,\Delta}$, $\mathbf{s}_{T,\Delta}$, $\mathbf{n}_{F,\Delta}$ to another plane. The two planes intersect one another on the directional sphere at the end point of vector $\mathbf{p}$. The three coordinate systems required to prove the refraction law are shown on the right-hand side of Figure B-1. One should imagine them to be shifted parallel to their indicated axes over to the origin at $\mathbf{O}_T$. Note that the following condition holds true: $y_F = y_I = \bar{y}_T$. Let us introduce the following three infinitesimal vectors

$$\Delta\mathbf{n} = \mathbf{n}_{F,\Delta} - \mathbf{n}_F, \tag{B.1}$$

$$\Delta\mathbf{s}_I = \mathbf{s}_{I,\Delta} - \mathbf{s}_I, \tag{B.2}$$

and

$$\Delta\mathbf{s}_T = \mathbf{s}_{T,\Delta} - \mathbf{s}_T, \tag{B.3}$$

with the understanding that $\Delta\mathbf{n}$ is measured in the $[x_F, y_F, z_F]$ system, $\Delta\mathbf{s}_I$ in the $[x_I, y_I, z_I]$ system, and $\Delta\mathbf{s}_T$ in the $[\bar{x}_T, \bar{y}_T, \bar{z}_T]$ system.

With this understanding and given the quadratic surface approximation for wavefronts and interfaces, these differential vectors have negligible components in the z-directions of their respective systems. Neglecting second- and higher-order terms of the Δ-quantities, we therefore have $\Delta\mathbf{n} \approx (\Delta n_x, \Delta n_y, 0)$; $\Delta\mathbf{s}_I \approx (\Delta s_{I,x}, \Delta s_{I,y}, 0)$; $\Delta\mathbf{s}_T \approx (\Delta s_{T,x}, \Delta s_{T,y}, 0)$. More precisely we have $\Delta n_z = \mathbf{O}(|\Delta\mathbf{n}|^2)$;

$$\text{i.e., } \left|\frac{\Delta n_z}{|\Delta\mathbf{n}|^2}\right| < \text{const for } \Delta\mathbf{n} \rightarrow 0,$$

and corresponding behavior for $\Delta s_{I,z}$ and $\Delta s_{T,z}$. Δn_x is the component of $\Delta\mathbf{n}$ in the x_F-direction and Δn_y in the y_F-direction, etc.

Let us also introduce the following abbreviations: $\mathbf{r}_F = (x_F, y_F)$; $\Delta\mathbf{r}_F = (\Delta x_F, \Delta y_F)$; $\mathbf{r}_I = (x_I, y_I)$; $\Delta\mathbf{r}_I = (\Delta x_I, \Delta y_I)$; $\bar{\mathbf{r}}_T = (\bar{x}_T, \bar{y}_T)$; $\Delta\bar{\mathbf{r}}_T = (\Delta\bar{x}_T, \Delta\bar{y}_T)$. $\Delta\mathbf{r}_F$ is the vector pointing from $\mathbf{O}_T$ to the projection of $\mathbf{O}_{T,\Delta}$ into the $x_F - y_F$ plane. With the quadratic surface assumption, $\Delta\mathbf{r}_F$ is approximately equal to the vector from $\mathbf{O}_T$ to $\mathbf{O}_{T,\Delta}$; the difference being again $\mathbf{O}(|\Delta\mathbf{r}|^2)$. $\Delta\mathbf{r}_I$ and $\Delta\bar{\mathbf{r}}_T$ are, respectively, the vectors from $\mathbf{O}_T$ to the projections of $\mathbf{O}_{T,\Delta}$ onto the $x_I - y_I$ plane and $\bar{x}_T - \bar{y}_T$ plane.

One can express $\Delta\mathbf{n}$, $\Delta\mathbf{s}_I$, and $\Delta\mathbf{s}_T$ within their respective coordinate systems in terms of curvature matrices, again neglecting second- and higher-order terms of the Δ-quantities:

$$\Delta\mathbf{n} \approx -\frac{1}{2}\nabla\,\mathbf{r}_F\,\mathbf{B}\,\mathbf{r}_F^T|_{\Delta\mathbf{r}_F} = -\Delta\mathbf{r}_F\,\mathbf{B}, \tag{B.4}$$

$$\Delta\mathbf{s}_I \approx \frac{1}{2}\nabla\,\mathbf{r}_I\,\mathbf{A}_I\,\mathbf{r}_I^T|_{\Delta\mathbf{r}_I} = \Delta\mathbf{r}_I\,\mathbf{A}_I, \tag{B.5}$$

and

$$\Delta\mathbf{s}_T \approx \frac{1}{2}\nabla\,\bar{\mathbf{r}}_T\,\bar{\mathbf{A}}_T\,\bar{\mathbf{r}}_T^T|_{\Delta\bar{\mathbf{r}}_T} = \Delta\bar{\mathbf{r}}_T\,\bar{\mathbf{A}}_T. \tag{B.6}$$

Equation (B.4) is obtained as follows: The interface is approximated within the $[x_F, y_F, z_F]$ system as

$$f(x_F, y_F, z_F) \equiv z_F - \frac{1}{2}\mathbf{X}_F\,\mathbf{B}\,\mathbf{X}_F^T = 0. \tag{B.7}$$

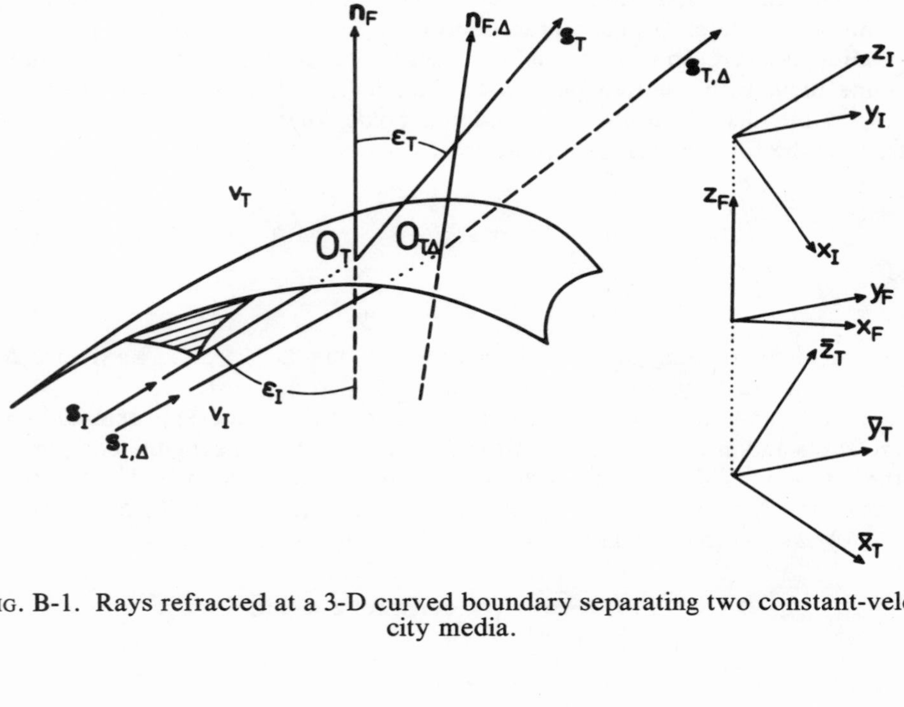

FIG. B-1. Rays refracted at a 3-D curved boundary separating two constant-velocity media.

The gradient vector of this surface at the coordinate origin is

$$\left(\frac{\partial f}{\partial x_F}, \frac{\partial f}{\partial y_F}, \frac{\partial f}{\partial z_F}\right)\Bigg|_{\mathbf{O}_T} = (0, 0, 1) = \mathbf{n}_F. \tag{B.8}$$

The gradient vector at point $\mathbf{O}_{T,\Delta}$ with coordinates $(\Delta x_F, \Delta y_F, \Delta z_F)$ is

$$\begin{aligned}\left(\frac{\partial f}{\partial x_F}, \frac{\partial f}{\partial y_F}, \frac{\partial f}{\partial z_F}\right)\Bigg|_{\mathbf{O}_{T,\Delta}} &\approx \mathbf{n}_{F,\Delta} \\ &= (-b_{11}\Delta x_F - b_{12}\Delta y_F, -b_{12}\Delta x_F - b_{22}\Delta y_F, 1).\end{aligned} \tag{B.9}$$

Subtracting $\mathbf{n}_F$ from $\mathbf{n}_{F,\Delta}$ provides

$$\Delta\mathbf{n} = \mathbf{n}_{F,\Delta} - \mathbf{n}_F \approx -\Delta\mathbf{r}_F\mathbf{B}. \tag{B.10}$$

In a similar way, one can establish equations (B.5) and (B.6).

By making use of the transformations (A.2) and (A.3), we can now project the vector $\Delta\mathbf{r}_F$ into the $x_I - y_I$ plane and the $\bar{x}_T - \bar{y}_T$ plane. This provides the two equations:

$$\Delta\mathbf{r}_I = \Delta\mathbf{r}_F\,\mathbf{S}_I, \tag{B.11}$$

and

$$\Delta\bar{\mathbf{r}}_T = \Delta\mathbf{r}_F\,\mathbf{S}_T, \tag{B.12}$$

where

$$\mathbf{S}_I = \begin{bmatrix}\cos\varepsilon_I & 0\\ 0 & 1\end{bmatrix}; \quad \mathbf{S}_T = \begin{bmatrix}\cos\varepsilon_T & 0\\ 0 & 1\end{bmatrix}.$$

With the help of the directional sphere, we will now show that the following relationship exists between the infinitesimal vectors $\Delta\mathbf{n}$, $\Delta\mathbf{s}_I$, and $\Delta\mathbf{s}_T$

$$\Delta\mathbf{s}_T \approx \frac{v_T}{v_I}\Delta\mathbf{s}_I\,\mathbf{S} - \rho\,\Delta\mathbf{n}\,\mathbf{S}_T^{-1}, \tag{B.13}$$

where

$$\rho = \frac{v_T}{v_I}\cos\varepsilon_I - \cos\varepsilon_T = \sin(\varepsilon_T - \varepsilon_I)/\sin\varepsilon_I, \tag{B.14}$$

and

$$\mathbf{S} = \mathbf{S}_I\,\mathbf{S}_T^{-1}.$$

The proof is as follows. Snell's law for the ray through $\mathbf{O}_T$ can be written as

$$\frac{\sin\varepsilon_T}{\sin\varepsilon_I} = \frac{\sin\langle\mathbf{n}_F, \mathbf{s}_T\rangle}{\sin\langle\mathbf{n}_F, \mathbf{s}_I\rangle} = \frac{v_T}{v_I}, \tag{B.15}$$

where $\langle\mathbf{n}_F, \mathbf{s}_I\rangle$ denotes the angle between the two unit vectors $\mathbf{n}_F$ and $\mathbf{s}_I$. For small angles $\Delta\alpha$, i.e., for angles which are $\mathbf{O}(|\mathbf{r}|)$ as for instance, $|\Delta\mathbf{n}|$, $|\Delta\mathbf{s}_I|$, and $|\Delta\mathbf{s}_T|$ on account of equations (B.4)–(B.6), we will use $\cos(\Delta\alpha) \approx 1$ and $\tan(\Delta\alpha) \approx \sin(\Delta\alpha) \approx \Delta\alpha$, neglecting quantities which are $\mathbf{O}(\Delta\alpha^2)$.

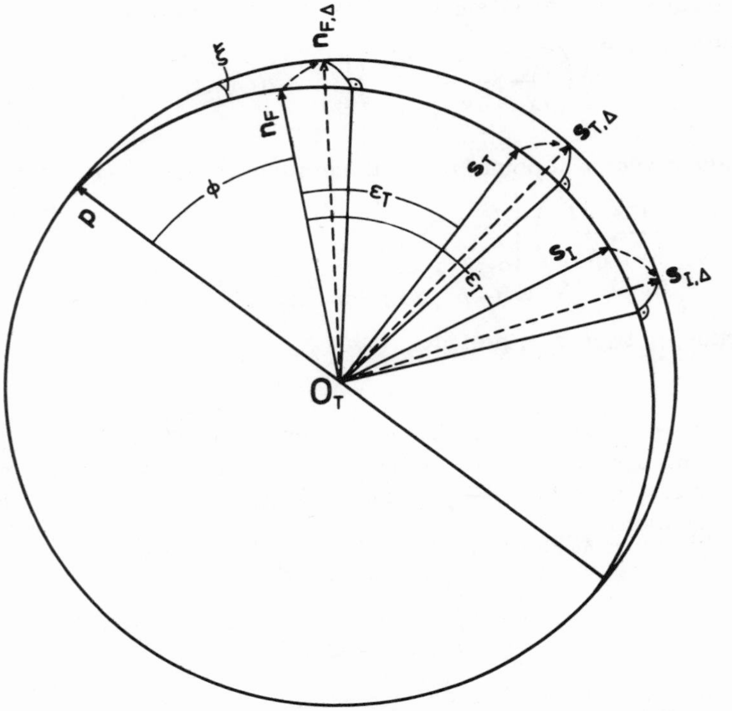

FIG. B-2. Directional sphere featuring directional ray vectors and interface normal vectors.

More precisely, in equations (B.16) to (B.18), all terms which are $\mathbf{O}(\Delta\alpha^2)$ are omitted. Neper's rule for spherical triangles on the directional sphere (Figure B-2) leads to the following approximation for Snell's law for the offset ray

$$\frac{v_T}{v_I} = \frac{\sin \langle \mathbf{n}_{F,\Delta}, \mathbf{s}_{T,\Delta} \rangle}{\sin \langle \mathbf{n}_{F,\Delta}, \mathbf{s}_{I,\Delta} \rangle} \approx \frac{\sin(\varepsilon_T - \Delta n_x + \Delta s_{T,x})}{\sin(\varepsilon_I - \Delta n_x + \Delta s_{I,x})}. \tag{B.16}$$

Using the addition law for the sine function and considering the properties of small angles changes equation (B.16) into

$$\Delta s_{T,x} \approx \frac{v_T \cos \varepsilon_I}{v_I \cos \varepsilon_T} \Delta s_{I,x} + \Delta n_x \frac{1}{\cos \varepsilon_T} \left(\cos \varepsilon_T - \frac{v_T}{v_I} \cos \varepsilon_I \right). \tag{B.17}$$

ϕ is the spherical distance along the great circle from $\mathbf{p}$ to $\mathbf{n}_F$. ξ is the angle between the two great circles of Figure B-2.

Neper's rules provide the following conditions

$$\cot \xi \tan(\Delta n_y) \approx \sin(\phi + \Delta n_x),$$
$$\cot \xi \tan(\Delta s_{I,y}) \approx \sin(\phi + \varepsilon_I + \Delta s_{I,x}),$$

and

$$\cot \xi \tan (\Delta s_{T,y}) \approx \sin (\phi + \varepsilon_T + \Delta s_{T,x}).$$

Eliminating ξ results in

$$\Delta n_y \approx \frac{\sin \phi}{\sin (\phi + \varepsilon_I)} \Delta s_{I,y},$$

and

$$\Delta n_y \approx \frac{\sin \phi}{\sin (\phi + \varepsilon_T)} \Delta s_{T,y}.$$

Again applying the addition law to the sine function leads to

$$\Delta s_{I,y} \approx \Delta n_y \cos \varepsilon_I + \Delta n_y \sin \varepsilon_I \cot \phi,$$

and

$$\Delta s_{T,y} \approx \Delta n_y \cos \varepsilon_T + \Delta n_y \sin \varepsilon_T \cot \phi.$$

Eliminating ϕ and using $\sin \varepsilon_T / \sin \varepsilon_I = v_T / v_I$ finally yields

$$\Delta s_{T,y} \approx \frac{v_T}{v_I} \Delta s_{I,y} + \Delta n_y \left(\cos \varepsilon_T - \frac{v_T}{v_I} \cos \varepsilon_I \right). \tag{B.18}$$

Combining equations (B.17) and (B.18) results in equation (B.13).

Substituting equations (B.4) to (B.6) and (B.11) and (B.12) into equation (B.13) leads, in the limit where the offset ray through $\mathbf{O}_{T,\Delta}$ approaches the ray through $\mathbf{O}_T$, from two different directions, e.g., from the x_F and y_F direction, to the following formula

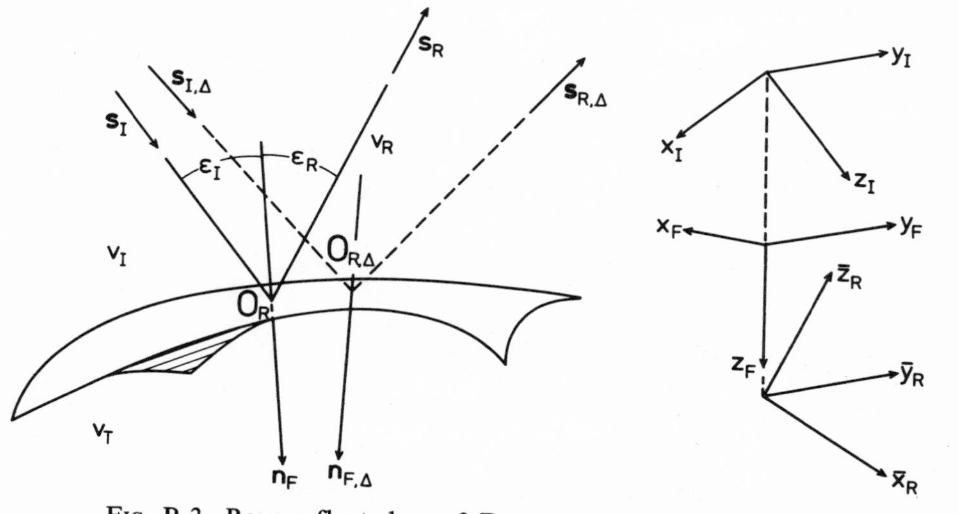

FIG. B-3. Rays reflected at a 3-D curved velocity boundary.

$$\bar{\mathbf{A}}_T = \frac{v_T}{v_I} \mathbf{S}\,\mathbf{A}_I\,\mathbf{S} + \rho\,\mathbf{S}_T^{-1}\,\mathbf{B}\,\mathbf{S}_T^{-1}. \tag{B.19}$$

As mentioned above, $\bar{\mathbf{A}}_T$ is the wavefront curvature matrix of the transmitted wave with respect to the $[\bar{x}_T, \bar{y}_T, \bar{z}_T]$ system.

Expressing the transmitted matrix within the $[x_T, y_T, z_T]$ system [see (A.5) to (A.7)] leads to

$$\mathbf{A}_T = \mathbf{D}^{-1}\,\bar{\mathbf{A}}_T\,\mathbf{D},$$

or

$$\mathbf{A}_T = \mathbf{D}^{-1}\left(\frac{v_T}{v_I} \mathbf{S}\,\mathbf{A}_I\,\mathbf{S} + \rho\,\mathbf{S}_T^{-1}\,\mathbf{B}\,\mathbf{S}_T^{-1}\right)\mathbf{D}. \tag{B.20}$$

Equation (B.20) is the refraction law for wavefront curvatures.

Let us finally make some remarks with regard to the *reflection law of wavefront curvature.* Figure B-3 shows a reflecting interface and two reflection points $\mathbf{O}_R$ and $\mathbf{O}_{R,\Delta}$ for two close rays. $\mathbf{s}_I$ and $\mathbf{s}_R$ are unit vectors within the incident and reflected ray at $\mathbf{O}_R$. $\mathbf{s}_{I,\Delta}$ and $\mathbf{s}_{R,\Delta}$ are their counterparts in the ray through $\mathbf{O}_{R,\Delta}$. For the three coordinate systems used at $\mathbf{O}_R$, the following condition holds true: $y_I = y_F = \bar{y}_R$. By using the directional sphere again, one can prove the reflection law by pursuing steps similar to those used to prove the refraction law. We avoid repeating these steps here; one can readily conclude that the resulting reflection law is closely related to the refraction law.

If no special significance is attributed to the difference between reflected or refracted rays, one can see that Figures B-1 and B-2 are identical when ε_R and v_R are replaced by $\varepsilon_T - \pi$ and v_T, respectively. This simple fact is revealed again in the law. It is obtained from the refraction law by replacing v_T with v_R and ε_T with $\varepsilon_R - \pi$. Compare also equation (A.3) with equation (A.4).

Appendix C
Derivation of the transmission law of divergence

Equation (4.73) is obtained with the help of Figure 4-15. The vector $\Delta\mathbf{r}_A$ points from ray point $\mathbf{P}_A$ to some point on the margin of the small elliptical area ΔF_A, a portion of the wavefront through $\mathbf{P}_A$. A ray through the endpoint of $\Delta\mathbf{r}_A$ passes through that of $\Delta\mathbf{r}_B$ of the advanced wavefront at $\mathbf{P}_B$. The wavefront approximation at $\mathbf{P}_A$ within a principal $[x', y', z']$ coordinate system (i.e., the x' and y' axes are pointing in the directions of extremal curvatures and the z'-axis in the direction of the ray) is

$$2z' = -\mathbf{X}'\,\mathbf{R}_A'^{-1}\,\mathbf{X}'^T, \tag{C.1}$$

where

$$\mathbf{R}_A'^{-1} = \begin{bmatrix} 1/R'_{A,x} & 0 \\ 0 & 1/R'_{A,y} \end{bmatrix},$$

and $R'_{A,x}$, $R'_{A,y}$ are the principal radii of wavefront curvature.

The corresponding approximation at $\mathbf{P}_B$ is

$$2z' = -\mathbf{X}'\,\mathbf{R}_B'^{-1}\,\mathbf{X}'^T. \tag{C.2}$$

With $\Delta\mathbf{a}$ and $\Delta\mathbf{b}$ being the vectors of the ellipse in the direction of the principal axes, the margin of the area ΔF_A is described by

$$\Delta\mathbf{r}_A(\phi) = (\Delta\mathbf{a}\cos\phi + \Delta\mathbf{b}\sin\phi), \tag{C.3}$$

so that the area itself is $2\pi|\Delta\mathbf{a}|\,|\Delta\mathbf{b}|$. The wavefront normal vector at $\mathbf{P}_A$ is $\mathbf{n}_0 = (0, 0, 1)$ and at $\Delta\mathbf{r}_A$, it is

$$\mathbf{n}_A \approx \frac{1}{2}\,\nabla\,(2z + \mathbf{X}'\,\mathbf{R}_A'^{-1}\,\mathbf{X}'^T)|_{\mathbf{X}'=\Delta\mathbf{r}_A} = (\Delta\mathbf{r}_A\,\mathbf{R}_A'^{-1}, 1), \tag{C.4}$$

where terms of the order $\mathbf{O}(|\Delta\mathbf{r}_A|^2)$ are neglected.

The vector $\Delta\mathbf{r}_B$ can consequently be obtained from

$$\begin{aligned}\Delta\mathbf{r}_B &\approx \Delta\mathbf{r}_A + s\,(\mathbf{n}_A - \mathbf{n}_0),\\ &\approx \Delta\mathbf{r}_A\,(\mathbf{I} + s\,\mathbf{R}_A'^{-1}),\end{aligned} \tag{C.5}$$

where s is the distance between $\mathbf{P}_A$ and $\mathbf{P}_B$.

The elliptical area ΔF_B is

$$\Delta F_B \approx 2\pi|\Delta\mathbf{a}|\,|\Delta\mathbf{b}|(1 + s/R'_{A,x})(1 + s/R'_{A,y});$$

consequently, when $\Delta F_A \to dF_A$ and $\Delta F_B \to dF_B$,

$$\frac{dF_B}{dF_A} = \frac{(R'_{A,x} + s)(R'_{A,y} + s)}{R'_{A,x} \cdot R'_{A,y}} = \frac{R'_{B,x}\, R'_{B,y}}{R'_{A,x}\, R'_{A,y}} = \frac{\det \mathbf{R}'_B}{\det \mathbf{R}'_A}. \tag{C.6}$$

As the determinant is invariant to a rotation of the coordinate system, one can also generally write

$$\frac{dF_B}{dF_A} = \frac{\det \mathbf{R}_B}{\det \mathbf{R}_A},$$

where $\mathbf{R}_A$ and $\mathbf{R}_B$ may refer to any right-handed coordinate system (not necessarily the principal or the moving coordinate system) at $\mathbf{P}_A$ or $\mathbf{P}_B$ that has its z-axis parallel to the selected ray.

Appendix D
The NIP wavefront approximation

Let us assume we have a 3-D isovelocity layer model (Figure D-1), a CDP profile, and a primary (or symmetric multiple) reflection. $S(r)$ is a source point and $G(r)$ is a receiver. Let $D(r)$ be the reflection point belonging to the source-receiver vector $\mathbf{r}$ of length r, and let $\mathbf{l}(r)$ be the vector of length $l(r)$ pointing from $D(r)$ to $D(0)$, where $D(0)$ is the normal incidence point NIP. l must be an even function of r. Thus, if interfaces are analytic functions and no focusing of CDP rays occurs at the CDP profile, we may write

$$l(r) = \text{const} \cdot r^2 + \mathbf{O}(r^4) = \mathbf{O}(r^2), \tag{D.1}$$

because $\mathbf{l}(r)$ is also analytic in this case and because $\mathbf{l}(0) = 0$.

Equation (D.1) can in fact be derived by recursion (starting from the one-layer case) without requesting interfaces to be analytic (Krey, 1976, 100–101).

Let us indicate this possibility here briefly prior to making use of equation (D.1) for CDP traveltime considerations. Let $\varepsilon = \varepsilon(r)$ be the reflection angle at $D(r)$; i.e., half the angle between the upgoing ray $R_u(r)$ and the downgoing ray $R_d(r)$ at $D(r)$. Let us also consider the normal incidence ray $R_0(r)$ at $D(r)$ which emerges at the surface at $0_0(r)$. It intersects the ith interface at $O_i(r)$, while the corresponding intersections at this interface for the up- and downgoing rays are $G_i(r)$ and $S_i(r)$. The angle which turns the direction of the ray $R_0(r)$ into that of $R_u(r)$ in the ith layer is called $\varepsilon_{u,i}$, the corresponding angle for $R_0(r)$ and $R_d(r)$ is $\varepsilon_{d,i}$.

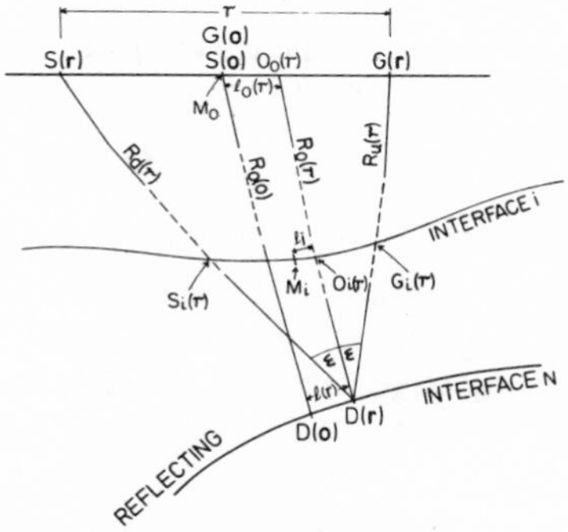

FIG. D-1. Normal ray and CDP ray for offset r.

It can be shown that the two equations

$$\varepsilon_{u,i} + \varepsilon_{d,i} = \mathbf{O}(\varepsilon^2), \tag{D.2}$$

and

$$|\mathbf{l}_i(r)| = \mathbf{O}(\varepsilon^2), \tag{D.3}$$

can be obtained by recursion starting with the deepest (i.e., the Nth) layer. In equation (D.3) $\mathbf{l}_i(r)$ is the vector from $O_i(r)$ to the midpoint M_i between S_i and G_i.

Both equations can easily be obtained for 2-D plane-dipping isovelocity layers by using the first two terms of the Taylor series expansions of the cosine and sine of $\varepsilon_{u,i}$ and $\varepsilon_{d,i}$ and some simple geometry. The extension to three dimensions is also of no fundamental difficulty, but $\varepsilon_{u,i}$ and $\varepsilon_{d,i}$ are then arcs on the directional sphere. Admitting interface curvatures is also allowed because curvature causes equal alterations of opposite sign of the angles $\varepsilon_{u,i}$ and $\varepsilon_{d,i}$ if terms of $\mathbf{O}(\varepsilon^2)$ are neglected. This is true because the vectors $\overrightarrow{O_i(r)\, G_i(r)}$ and $\overrightarrow{O_i(r)\, S_i(r)}$ are of equal length but of opposite signs if terms of $\mathbf{O}(\varepsilon^2)$ are neglected. Note that the vectors themselves are $\mathbf{O}(\varepsilon)$.

Thus, at the surface of the earth model

$$l_0(r) = |\overrightarrow{M_0 O_0(r)}| = \mathbf{O}(\varepsilon^2) = \mathbf{O}(r^2). \tag{D.4}$$

Now the distances $l_i(r)$ between the two normal incidence rays $R_0(0)$ and $R_0(r)$ (emerging at M_0 [$=G(0) = S(0)$] and $\mathbf{O}_0(r)$, respectively) can be estimated at each interface i. By recursion from the surface down to the reflecting interface N we obtain.

$$l(r) = |\overrightarrow{D(0)\, D(r)}| = \mathbf{O}(r^2),$$

provided no focusing occurs at M_0.

Let us now use this result to demonstrate that the NIP wavefront represent the CDP traveltime curve up to the second order in $|\mathbf{r}|$.

Let us consider the two wavefronts through $D(r)$ that are assumed to have originated at the point source $S(r)$ and the receiver point $G(r)$. The difference between traveltime of the ray from $S(r)$ to $D(r)$ and that from $S(r)$ to $D(0)$ is

$$\Delta t_{S,r} = \mathbf{l}(r) \cdot \nabla t_{S,r} + \mathbf{O}[l^2(r)], \tag{D.5}$$

where $t_{S,r}$ is the traveltime from $S(r)$ to any point in the vicinity of $D(r)$, and where $\nabla t_{S,r}$ is the gradient of this traveltime at $D(r)$. The corresponding equation referring to $G(r)$ instead of $S(r)$ is

$$\Delta t_{G,r} = \mathbf{l}(r) \cdot \nabla t_{G,r} + \mathbf{O}[l^2(r)]. \tag{D.6}$$

According to the reflection law, the components of the gradient vectors in the expressions (D.5) and (D.6) which lie in the plane tangential to the reflecting interface at $D(r)$ are equal but of opposite sign. The component of $\mathbf{l}(r)$ perpendicular to the tangential plane is $\mathbf{O}[l^2(r)]$ because $D(0)$ is the NIP.

By using equations (D.1), (D.5), and (D.6), we find that the total difference Δt_r of the raypaths from $S(r)$ via $D(r)$ to $G(r)$ on the one hand and from $S(r)$ via $D(0)$ to $G(r)$ on the other hand, therefore, becomes

$$\Delta t_r = \Delta t_{S,r} + \Delta t_{G,r} = \mathbf{O}[l^2(r)] = \mathbf{O}(r^4), \tag{D.7}$$

and can thus be neglected when $|\mathbf{r}|$ is small.

The concept of the NIP wavefront thus readily holds and can consequently be employed to describe CDP traveltime functions in the vicinity of $S(0)$ including terms of the second order in $|\mathbf{r}|$.

Appendix E
Transmission law of wavefront curvature

Theorem

Given a wavefront at time zero with principal radii of curvature R'_x and R'_y in a medium of constant velocity v, the principal radii of curvature of the advanced wavefront at time Δt are $R'_x + v\Delta t$ and $R'_y + v\Delta t$.

Proof

In the vicinity of the origin of a principal $[x'_0, y'_0, z'_0]$ coordinate system, the wavefront at time zero may be represented as

$$z'_0 + \frac{x'^2_0}{2R'_x} + \frac{y'^2_0}{2R'_y} = 0, \tag{E.1}$$

where the direction of the central ray is the z-axis. We have to prove that, in the vicinity of the z-axis, the advanced wavefront at time Δt is

$$z' - v\Delta t + \frac{x'^2}{2(R'_x + v\Delta t)} + \frac{y'^2}{2(R'_y + v\Delta t)} = 0. \tag{E.2}$$

That is, we have to show that the minimum distance d of any point on surface (E.1) from surface (E.2) is everywhere equal to $v\Delta t$. The square of the distance is given by $d^2 = (x' - x'_0)^2 + (y' - y'_0)^2 + (z' - z'_0)^2$. Subtracting equation (E.1) from equation (E.2) yields

$$z' - z'_0 = v\Delta t + \frac{1}{2}\left(\frac{x'^2_0}{R'_x} - \frac{x'^2}{R'_x + v\Delta t} + \frac{y'^2_0}{R'_y} - \frac{y'^2}{R'_y + v\Delta t}\right), \tag{E.3}$$

and neglecting fourth and higher powers of x', x'_0, y', and y'_0, we can write

$$d^2 = (v\Delta t)^2 + (v\Delta t)\left(\frac{x'^2_0}{R'_x} - \frac{x'^2}{R'_x + v\Delta t} + \frac{y'^2_0}{R'_y} - \frac{y'^2}{R'_y + v\Delta t}\right) + (x'_0 - x')^2 + (y'_0 - y')^2, \tag{E.4}$$

from which follows that

$$\frac{\partial d^2}{\partial x'_0} = 2\left[\left(\frac{v\Delta t}{R'_x} + 1\right) x'_0 - x'\right] = 0, \tag{E.5}$$

for

$$x'_{0,\min} = \left(\frac{v\Delta t}{R'_x} + 1\right)^{-1} x'. \tag{E.6}$$

At this point, the extremal value of d^2 is a minimum since $\partial^2 d^2/\partial x'^2 > 0$.

Similarly, we get

$$y'_{0,\min} = \left(\frac{v\,\Delta t}{R'_y} + 1\right)^{-1} y'. \tag{E.7}$$

Substituting equations (E.6) and (E.7) into equation (E.4) yields

$$d^2\,(x'_{0,\min}, y'_{0,\min}) = (v\,\Delta t)^2. \tag{E.8}$$

Thus $(v\,\Delta t)^2$ is not only a minimum value, but it is also obtained at a particular (x'_0, y'_0) point relative to (x', y').

In normal sections that are oblique to the principal directions, the radii generally do not increase linearly with distance. More precisely, if the curve formed by the intersection of the wavefront approximation (E.1) with the plane $y' = x' \tan \phi$ has the radius of curvature $R'_{0,\phi}$, then the corresponding radius of curvature $R'_{\Delta t,\phi}$ of a curve formed by the intersection of equation (E.2) with plane $y' = x' \tan \phi$ is not equal to $R'_{0,\phi} + v\Delta t$. Thus $R'_{\Delta t,\phi} = R'_{0,\phi} + v\Delta t$ *only* when $\phi = 0$, or $\pi/2 \bmod \pi$ that is, when $R'_x = R'_y$. In fact, it is clear from inspection that, in general,

$$R'_{\Delta t,\phi} = \frac{1}{\dfrac{\cos^2\phi}{R'_x + v\,\Delta t} + \dfrac{\sin^2\phi}{R'_y + v\,\Delta t}}.$$

Appendix F
Traveltime inversion for a medium with a linear vertical velocity gradient

Assuming a velocity law of the form $v(z) = v_0 + gz$ above a dipping reflector, we will now show how to determine v_0 and g from $V_{\mathrm{NMO},x}$, $V_{\mathrm{NMO},y}$, t_0, and ∇t_0 at CDP location 0_0 (Figure 9-6). We shall first provide a proof for equation (9.27), then find an analytic expression for $V_{\mathrm{NMO},y}$ with which we will finally obtain equation (9.29).

The CDP reflection time in the vicinity of 0_0 as a function of offset r in the x-direction can be expanded as

$$t(r) = t(0) + \Delta t_{\mathrm{NMO},x} + \ldots,$$

or

$$t(r) = t(0) + r^2/(2t_0 V^2_{\mathrm{NMO},x}) + \ldots.$$

Suppose source and receiver are placed symmetrically with respect to 0_0 along the x-axis. We can then express the NMO time in the x-direction as $\Delta t_{\mathrm{NMO},x} \approx (t_{\mathrm{NIP},S} - t_0/2) + (t_{\mathrm{NIP},G} - t_0/2)$. $t_{\mathrm{NIP},S}$ and $t_{\mathrm{NIP},G}$ are the traveltimes consumed by the NIP wave between the NIP and S and the NIP and G, respectively. If R_{NIP} is the radius of curvature of the emerging NIP wavefront, we can write

$$\begin{aligned}\Delta t_{\mathrm{NMO},x} = \frac{r}{2}\sin\beta_0 &\left[\left(v_0 + \frac{1}{4} g\, r \sin\beta_0 \cos\beta_0\right)^{-1}\right.\\ &\left.- \left(v_0 - \frac{1}{4} g\, r \sin\beta_0 \cos\beta_0\right)^{-1}\right]\\ &+ \left(\frac{r}{2}\right)^2 \cos^2\beta_0/(R_{\mathrm{NIP}}\, v_0) + \mathbf{O}(r^4).\end{aligned} \tag{F.1}$$

The second term on the right side of equation (F.1) refers to the ray segment between the emerging NIP wavefront and its tangential plane at 0_0. The first term refers to the ray segments between the tangential plane on the one hand and shotpoint and geophone point on the other hand. The traveltimes through these latter two segments are of opposite sign. The sum would be zero in the case of homogeneity. In our case, however, the first term of equation (F.1) results.

In order to derive this term, we replace the ray segments by straight lines which are perpendicular to the tangential plane, still passing through shot point and geophone point, respectively. The resulting time difference is $\mathbf{O}(r^3)$, as can be easily derived. The time consumed for these straight segments can be computed by dividing the lengths $\pm\, r/2\ (\sin\beta_0)$ of the segments by the vertical average velocity $v_a(z^*)$ between $z = 0$ and $z^* = \pm\, r/2\ (\sin\beta_0)\cos\beta_0$, z^* being the elevation of the endpoints of the segments at the tangential plane. The average velocity $v_a(z^*)$ results from equation (9.25) by the well-known formula

$$v_a(z^*) = z^* \Big/ \int_0^{z^*} \frac{dz}{v(z)}$$
$$= v_0 + \frac{1}{2} g z^* + \mathbf{O}[(z^*)^2] .$$

With $z^* = \mathbf{O}(r)$, and taking account of the fact that $\Delta t_{\mathrm{NMO},x}$ is an even analytic function of r, and implying that $\mathbf{O}(r^3)$ can be replaced by $\mathbf{O}(r^4)$, equation (F.1) can now immediately be obtained.

By expanding equation (F.1) into a Taylor series in r and accounting for the fact that the NIP wavefront is a sphere, one can write

$$\Delta t_{\mathrm{NMO},x} = -\frac{1}{4} r^2 g \sin^2\beta_0 \cos\beta_0 / v_0^2 + \cos^2\beta_0\, \Delta t_{\mathrm{NMO},y} + \mathbf{O}(r^4).$$

By dividing both sides of this equation by r^2 and multiplying with $2t_0$, one obtains the following equation in the limit for $r \rightarrow 0$,

$$(\cos^2\beta_0)\, V_{\mathrm{NMO},y}^{-2} - V_{\mathrm{NMO},x}^{-2} = +g\, t_0 \sin^2\beta_0 \cos\beta_0 / 2v_0^2,$$

from which we find, by using equation (9.26), the desired result

$$(\cos\beta_0)\, V_{\mathrm{NMO},y}^{-2} - (\cos\beta_0)^{-1}\, V_{\mathrm{NMO},x}^{-2} = \frac{1}{8} g\, t_0 \left(\frac{\partial t_0}{\partial x}\right)^2 . \qquad \text{(F.2)}$$

Hubral (1979c) discusses, in some detail, the subject of relating NMO velocity to the NIP wavefront emerging from an *arbitrary inhomogeneous* medium.

Before finding an expression for $V_{\mathrm{NMO},y}$, let us review some fundamental properties of rays in a medium having a linear vertical velocity gradient.

Taking advantage of the fact that the normal ray is a circle (Michaels, 1977) with center point at the elevation (see Figure 9-6),

$$z_c = -v_0/g, \qquad \text{(F.3)}$$

we can introduce the angle β between the ray and the z-direction as an independent variable. $\beta(t_0)$ is the angle at the reflection point NIP. The radius R of the ray circle is given by

$$R = v_0/(g \sin\beta_0) = v/(g \sin\beta). \qquad \text{(F.4)}$$

With the origin of the coordinate system placed at 0_0 and recognizing that $Rd\beta$ is the increment ds of raypath length, we have the following equations for the ray

$$z\,[\beta(t)] = \int_{\beta_0}^{\beta(t)} ds \cos\beta = \int_{\beta_0}^{\beta(t)} R d\beta \cos\beta$$
$$= R\,[\sin\beta(t) - \sin\beta_0], \qquad \text{(F.5)}$$

and

$$t\,[\beta(t)] = 2\,R \int_{\beta_0}^{\beta(t)} d\beta / v(z) = 2 \int_{\beta_0}^{\beta(t)} d\beta / g \sin\beta, \tag{F.6}$$

where t is the two-way time from 0_0 to ray point (z, t).

Carrying out the integration in equation (F.6) and evaluating the result when $t = t_0$, yields

$$t_0\,[\beta(t_0)] = \frac{2}{g} \ln \{\tan\,[\beta(t_0)/2]\,/\tan(\beta_0/2)\}. \tag{F.7}$$

In the y-direction, the Dix formula can be applied to the nonvertical normal ray. This fact can be deduced, for instance, from considering equations (4.85) and (6.12). Applying equation (F.4) and the linear velocity law, we can write Dix's formula as

$$\begin{aligned} t_0\,V^2_{\mathrm{NMO},y} &= \int_0^{t_0} dt\; v^2(t) = 2 \int_0^{S} ds\; v(z) = 2 \int_{\beta_0}^{\beta(t_0)} R\, d\beta\; v(z) \\ &= 2R \int_{\beta_0}^{\beta(t_0)} d\beta\; g\,R \sin\beta. \end{aligned}$$

Integration results in

$$t_0\,V^2_{\mathrm{NMO},y} = [2\,v_0^2/(g \sin^2\beta_0)] \cdot \{\cos\beta_0 - \cos\,[\beta(t_0)]\}. \tag{F.8}$$

Dividing by equation (F.7) eliminates g; i.e.,

$$\begin{aligned} V^2_{\mathrm{NMO},y} &= v_0^2\,\{\cos\beta_0 - \cos\,[\beta(t_0)]\}/(\sin^2\beta_0). \\ &\quad \cdot \ln\{\tan\,[\beta(t_0)/2]\,/\tan(\beta_0/2)\}). \end{aligned} \tag{F.9}$$

Then by applying

$$v_0 = 2 \sin\beta_0 \Big/ \frac{\partial t_0}{\partial x},$$

we get

$$V^{-2}_{\mathrm{NMO},y} = \frac{1}{4}\left(\frac{\partial t_0}{\partial x}\right)^2 \ln\{\tan\,[\beta(t_0)/2]\,/\tan(\beta_0/2)\}/\{\cos\beta_0 - \cos\,[\beta(t_0)]\}. \tag{F.10}$$

Introducing equation (F.7) into equation (F.2) and utilizing equation (F.10), yields

$$\begin{aligned} \cos\beta_0\,V^{-2}_{\mathrm{NMO},y} &- (\cos\beta_0)^{-1}\,V^{-2}_{\mathrm{NMO},x} \\ &= \frac{1}{4}\left(\frac{\partial t_0}{\partial x}\right)^2 \ln\{\tan\,[\beta(t_0)/2]\,/\tan(\beta_0/2)\} \\ &= V^{-2}_{\mathrm{NMO},y}\,[\cos\beta_0 - \cos\beta(t_0)], \end{aligned} \tag{F.11}$$

or

$$\cos\beta(t_0)\cos\beta_0 = V^2_{\mathrm{NMO},y}/V^2_{\mathrm{NMO},x}. \tag{F.12}$$

Solving for $\beta(t_0)$ in both equations (F.11) and (F.12) results in

$$\beta(t_0) = 2 \arctan \left\{ \tan(\beta_0/2) \exp \left[4 \left(\frac{\partial t_0}{\partial x} \right)^{-2} \right.\right.$$

$$\left.\left. [\cos\beta_0 \, V_{\mathrm{NMO},y}^{-2} - (\cos\beta_0)^{-1} V_{\mathrm{NMO},x}^{-2}] \right] \right\},$$

$$= \arccos \left[(\cos\beta_0)^{-1} V_{\mathrm{NMO},y}^{2} V_{\mathrm{NMO},x}^{-2} \right]. \tag{F.13}$$

This equation has β_0 as its only unknown; it can be determined by trial and error. Once it is determined, we can compute $\beta(t_0)$, v_0, and g, as well as the location of the NIP and the dip of the reflector.

References

Ahlūwalia, D. S., and Keller, J. B., 1977, Exact and asymptotic representations of a sound field in a stratified ocean, *in* Wave propagation and underwater acoustics: J. B. Keller and J. S. Papadakis, Eds., Berlin, Lecture Notes in Physics 70, Springer-Verlag.

Ahlberg, J. H., Nilson, E. N., and Walsh, J. L., 1967, The theory of splines and their applications: New York, Academic Press.

Al-Chalabi, M., 1973, Series approximation in velocity and traveltime computations: Geophys. Prosp., v. 21, p. 783–795.

——— 1974, An analysis of stacking, RMS, average, and interval velocities over a horizontally layered ground: Geophys. Prosp., v. 22, p. 458–475.

Anstey, N. 1977, Seismic interpretation: The physical aspects: Int. Human Res. Dev. Corp., Boston.

Backus, M. M., and Chen, R. L., 1975, Flat spot exploration: Geophys. Prosp., v. 23, p. 533–577.

Bading, R., and Krey, Th., 1976, Optimum attenuation of multiples by appropriate CRP-field techniques and migrations: Presented at the 46th Annual International SEG Meeting, October 28, in Houston.

Båth, M., 1968, Mathematical aspects of seismology: Amsterdam, Elsevier.

Beitzel, J. E., and Davis, J. M., 1974, A computer oriented velocity analysis interpretation technique: Geophysics, v. 39, p. 619–632.

Berkhout, A. J., and Van Wulften Palthe, P. W., 1979, Migration in terms of spatial deconvolution: Geophys. Prosp., v. 27, p. 261–291.

Boisse, S., 1978, Calculation of velocity from seismic reflection amplitude: Geophys. Prosp., v. 26, p. 163–174.

Bodoky, T., and Szeidovitz, Zs., 1972, The effect of normal correction errors on the stacking of common-depth point traces: Geophys. Inst. Roland Eötvös, Geophys. Trans., v. 20 (3–4), p. 47–57.

Booker, A. H., Linville, A. F., and Wason, C. B., 1976, Long wavelength static estimation: Geophysics, v. 41, p. 939–959.

Bortfeld, R., 1973, Comments on "Series approximation in velocity and traveltime computations": Geophys. Prosp., v. 21, p. 796–797.

Brekhovskikh, L. M., 1960, Waves in layered media: New York, Academic Press, Inc.

Brown, R. J. S., 1969, Normal moveout and velocity relations for flat and dipping beds and for long offsets: Geophysics, v. 34, p. 180–195.

Buchholtz, H., 1972, A note on signal distortion due to dynamic (NMO) corrections: Geophys. Prosp., v. 20, p. 395–402.

Cagniard, L., 1962, Reflection and refraction of progressive seismic waves: Trans. E. A. Flinn and C. H. Dix, New York, McGraw-Hill Book Co., Inc.

Červený, V., and Ravindra, R., 1971, Theory of seismic headwaves: Toronto, University of Toronto Press.

Červený, V., Molotkov, I. A., and Pšenčik, I., 1977, Ray method in seismology: Univerzita Karlova Praha.

Claerbout, J. F., 1971, Toward a unified theory of reflector mapping: Geophysics, v. 36, p. 467–481.

Claerbout, J. F., and Doherty, S. M., 1972, Downward continuation of moveout corrected seismograms: Geophysics, v. 37, p. 741–768.

——— 1976, Fundamentals of geophysical data processing: with applications to petroleum prospecting: New York, McGraw-Hill Book Co., Inc.

Chander, R., 1977, On tracing rays with specified end points: Geophys. Prosp., v. 25, p. 120–125.

Clark, S. P., 1966, Handbook of physical constants: GSA memoir 97.

Cressman, K. S., 1968, How velocity layering and steep dip affect CDP: Geophysics, v. 33, p. 399–411.

Davis, J. M., 1972, Interpretation of velocity spectra through an adaptive modeling strategy: Geophysics, v. 37, p. 953–962.

Deschamps, G. A., 1972, Ray techniques in electromagnetics: Proc. IEEE, v. 60, p. 1022–1035.
Dinstel, W. L., 1971, Velocity spectra and diffraction patterns: Geophysics, v. 36, p. 415–417.
Dix, C. H., 1955, Seismic velocities from surface measurements: Geophysics, v. 20, p. 68–86.
Dobrin, M. B., 1976, Introduction to geophysical prospecting: New York, McGraw-Hill Book Co., Inc.
Doherty, S. M., and Claerbout, J. F., 1976, Structure independent velocity estimation: Geophysics, v. 41, p. 850–881.
Dohr, G. P., and Stiller, P. K., 1975, Migration velocity determination—Part II: Applications: Geophysics, v. 40, p. 6–16.
Domenico, S. N., 1974, Effect of water saturation on seismic reflectivity of sand reservoirs encased in shale: Geophysics, v. 39, p. 759–769.
Domenico, S. N., and Crowe, C., 1978, Interpretational need for seismic parameter estimates: SEG Cont. ed. symp.
Donaldson, J. A., 1972, Pick validation of velstack velocities: Presented at the 42nd Annual International SEG Meeting, November 30, in Anaheim.
Dunkin, J. W., and Levin, F. K., 1973, Effect of normal moveout on a seismic pulse: Geophysics, v. 38, p. 635–642.
Dürbaum, H., 1953, Possibilities of constructing true raypaths in reflection seismic interpretation: Geophys. Prosp., v. 1, p. 125–139.
——— 1954, Zur Bestimmung von Wellengeschwindigkeiten aus reflexions-seismischen Messungen: Geophys. Prosp., v. 2, p. 151–167.
Everett, J. E., 1974, Obtaining interval velocities from stacking velocities when dipping horizons are included: Geophys. Prosp., v. 22, p. 122–142.
Ewing, W. M., Jardetzky, W. S., and Press, F., 1957, Elastic waves in layered media: New York, McGraw-Hill Book Co., Inc.
Faust, L. Y., 1951, Seismic velocity as a function of depth and geologic time: Geophysics, v. 16, p. 192–206.
French, W. S., 1974, Two-dimensional and three-dimensional migration of model-experiment reflection profiles: Geophysics, v. 39, p. 265–277.
——— 1975, Computer migration of oblique seismic reflection profiles: Geophysics, v. 40, p. 961–980.
Galbraith, J. N., Jr., and Wiggins, R. A., 1968, Characteristics of optimum multichannel stacking filters: Geophysics, v. 33, p. 36–48.
Gangi, A. F., and Sung, J. Y., 1976, Traveltime curves for reflections in dipping layers: Geophysics, v. 41, p. 425–440.
Gardner, G. H. F., Gardner, L. W., and Gregory, A. R., 1974a, Formation velocity and density—the diagnostic basics for stratigraphic traps: Geophysics, v. 39, p. 770–780.
Gardner, G. H. F., French, W. S., and Matzuk, T., 1974b, Elements of migration and velocity analysis: Geophysics, v. 39, p. 811–825.
Garotta, R., 1971, Selection of seismic picking based upon the dip, moveout, and amplitude of each event: Geophys. Prosp., v. 19, p. 357–370.
Garotta, R., and Michon, D., 1967, Continuous analysis of the velocity function and of the moveout corrections: Geophys. Prosp., v. 15, p. 584–597.
Gassmann, F., 1951, Elastic waves through a packing of spheres: Geophysics, v. 16, p. 673–685.
——— 1964, Introduction to seismic traveltime methods in anisotropic media: Pure and Appl. Geophys., v. 58, p. 63–113.
——— 1972, Seismische Prospektion: Basel-Stuttgart, Birkhäuser Verlag.
Gel'chinskiy, B. Y., 1961, An expression for the spreading function, *in* Problems in the dynamic theory of propagation of seismic waves: G. I. Petrashen, Ed., Leningrad, Leningrad University Press, v. 5; p. 47–53.
Gerritsma, P. H. A., 1977, Time-to-depth conversion in the presence of structure: Geophysics, v. 42, p. 760–772.
Gibson, B. S., Odegard, M. E., and Sutton, G. H., 1979, Nonlinear least-squares inversion of traveltime data for a linear velocity-depth relationship: Geophysics, v. 44, p. 185–194.
Gjoystdal, H., and Ursin, B., 1978, Inversion of reflection in three dimensions: Presented at the 48th Annual International SEG Meeting, November 1, in San Francisco.
Gol'din, S. V., and Cernjak, V. S., 1976, Analogues of Dix's formulas for laminar homogeneous media with non-horizontal boundaries: Geol. and Geophys., no. 10 (in Russian).
Grant, F. S., and West, G. F., 1965, Interpretation theory in applied geophysics: New York, McGraw-Hill Book Co., Inc.

Green, C. H., 1938, Velocity determination by means of reflection profiles: Geophysics, v. 3, p. 295–305.

Gullstrand, A., 1915, Das allgemeine optische Abbildungssystem: Svenska Vetenskapsakademiens Handlingar, v. 55, p. 1–139.

Gutenberg, B., 1936, The amplitudes of waves to be expected in seismic prospecting: Geophysics, v. 1, p. 252–256.

Hagedoorn, J. G., 1954, A process of seismic reflection interpretation: Geophys. Prosp., v. 2, p. 85–127.

Hammond, A. L., 1974, Bright spot: Better seismological indicators of gas and oil: Science, v. 185, p. 515–517.

Hatton, L., Larner, K., and Gibson, B., 1978, Migration of seismic data from inhomogeneous media: Presented at the 48th Annual International SEG Meeting, October 30, in San Francisco.

Helbig, K., 1965, Die Indexfläche als Hilfsmittel für strahlengeometrische Konstruktionen bei der Interpretation seismischer Beobachtungen, insbesondere bei anisotropem Untergrund: Bayerische Akad. der Wissenschaften, math.-naturw. Klass, Abhandlungen, Neue Folge, Heft 122, München.

——— 1979, Discussion on "The reflection, refraction, and diffraction of waves in media with an elliptical velocity dependence" (Franklyn K. Levin): Geophysics, v. 44, p. 987–990.

Hilterman, F. J., 1970, Three-dimensional seismic modeling: Geophysics, v. 35, p. 1020–1037.

——— 1975, Amplitudes of seismic waves—A quick look: Geophysics, v. 40, p. 745–762.

Houba, W., and Krey, Th., 1976, An approach to 3-D migration from conventional or near-conventional line shooting: Presented at the 46th Annual International SEG Meeting October 27, in Houston.

Hosken, J. W. J., 1978, 2D migration by Kirchhoff summation: The penalties of cutting corners: Presented at the 40th EAEG Meeting, Dublin.

Hubral, P., 1974, Comments on "Obtaining interval velocities from stacking velocities when dipping horizons are included": Geophys. Prosp., v. 24, p. 719–724.

——— 1976a, Interval velocities from surface measurements in the three-dimensional plane layer case: Geophysics, v. 41, p. 233–242.

——— 1976b, CDP ray modeling in the presence of 3D plane isovelocity layers of varying dip and strike: Geophys. Prosp., v. 24, p. 478–491.

——— 1976c, Ray tracing and 3D seismic modeling: Paper presented at the 46th Annual International SEG Meeting October 28, in Houston.

——— 1977, Time migration—Some ray theoretical aspects: Geophys. Prosp., v. 25, p. 738–745.

——— 1979a, A wave front curvature approach to computing ray amplitudes in inhomogeneous media with curved interfaces: Studia Geoph. et Geod., v. 23, p. 131–137.

——— 1979b, A method of computing the NMO-velocity in laterally inhomogeneous media with curved interfaces: Geophys. Prosp., in press.

——— 1980, Wave front curvatures in 3D laterally inhomogeneous media with curved interfaces: Geophysics, v. 45, p. 905–913.

Judson, D., Lin, J., Schultz, P., and Sherwood, J., 1978, Depth migration after stack: Presented at the 48th Annual International SEG Meeting October 30, in San Francisco.

Julian, B. R., and Gubbins, D., 1977, Three-dimensional seismic ray tracing: J. of Geophys., v. 43, p. 95–113.

Kesmarky, I., 1976, Estimation of interval velocities and the geological model: Geophys. Inst. Roland Eötvös, Geophys. Trans., v. 24, p. 63–75.

——— 1977, Estimation of reflector parameters by the virtual image technique: Geophys. Prosp., v. 25, p. 621–635.

Kleyn, A. H., 1977, On the migration of reflection time contour maps: Geophys. Prosp., v. 25, p. 125–140.

Kline, M., 1961, A note on the expansion coefficient of geometrical optics: Comm. of Pure and Appl. Math., v. 14, p. 473–479.

Kline, M., and Kay, I. W., 1965, Electromagnetic theory and geometrical optics: New York, Interscience Publishers.

Krey, Th., 1951, An approximate correction method for refraction in reflection seismic prospecting: Geophysics, v. 16, p. 468–485.

——— 1954, Bemerkungen zu einer Formel für Geschwindigkeitsbestimmungen aus seismischen Messungen von C. H. Dix, Erdöl und Kohle: 7. Jahrg., Heft. 1, p. 8–9.

——— 1976, Computation of interval velocities from common reflection point moveout times for N layers with arbitrary dips and curvatures in three dimensions when assuming small shot-geophone distances: Geophys. Prosp., v. 24, p. 91–111.

——— 1978, Mapping non-reflecting velocity interfaces by normal moveout velocities of underlying horizons: Presented at the 40th EAEG Meeting, Dublin.
Krey, Th., and Helbig, K., 1956, A theorem concerning anisotropy of stratified media and its significance for reflection seismics: Geophys. Prosp., v. 3, p. 294–302.
Larner, K., and Rooney, M., 1972, Interval velocity computation for plane dipping multi-layered media: Paper presented at the 42nd Annual International SEG Meeting November 30, in Anaheim.
Larner, K., 1974, An overview of continuous velocity analysis: Cont. Ed. Sem. by the Geophysical Society of Houston, December 4.
Larner, K., and Hatton, L., 1976, Wave-equation migration: Two approaches: Presented at the 8th Annual Offshore Technology Conference, Houston.
Larner, K., Hatton, L., and Forshaw, R., 1977, Depth migration of imaged time sections: Presented at the 39th EAEG Meeting, Zagreb.
Lavergne, M., and Willm, C., 1977, Inversion of seismograms and pseudo-velocity logs: Geophys. Prosp., v. 25, p. 231–250.
Levin, F. K., 1971, Apparent velocity from dipping interface reflections: Geophysics, v. 36, p. 510–516.
——— 1979, Seismic velocities in transversely isotropic media: Geophysics, v. 44, p. 918–936.
Levin, F. K., and Shah, P. M., 1977, Peg-leg multiples and dipping reflectors: Geophysics, v. 42, p. 957–981.
Loewenthal, D., Lu, L., Roberson, R., and Sherwood, J., 1976, The wave equation applied to migration: Geophys. Prosp., v. 24, p. 380–399.
Lucas, A. L., Al-Chalabi, M., and Shaw, I. L., 1975, The calculation of laterally varying time delays from stacking velocity anomalies: Presented at the 45th Annual International SEG Meeting October 13, in Denver.
Marschall, R., 1975, Einige Probleme bei der Benutzung grosser Schuss-Geophon-Abstände und deren Anwendung auf Unterschiessungen: Ph. D. thesis, Leoben.
May, B. T., and Hron, F., 1978, Synthetic seismic sections of typical petroleum traps: Geophysics, v. 43, p. 1119–1147.
Mayne, W. H., 1962, Common reflection point horizontal data stacking technique: Geophysics, v. 27, p. 927–938.
——— 1967, Practical considerations in the use of common reflection point techniques: Geophysics, v. 32, p. 225–229.
Meixner, E., 1978, 3D determination of interval velocity: Paper presented at the 48th International SEG Meeting November 1, in San Francisco.
Meyerhoff, H. F., 1966, Horizontal stacking and multichannel filtering applied to common depth point seismic data: Geophys. Prosp., v. 14, p. 441–454.
Michaels, P., 1977, Seismic ray path migration with the pocket calculator: Geophysics, v. 42, p. 1056–1063.
Miller, M. K., 1974, Stacking of reflections from complex structures: Geophysics, v. 39, p. 427–440.
Montalbetti, J. F., 1971, Computer determination of seismic velocities—a review: Canadian SEG J., p. 32–41.
Müller, G., 1971, Direct inversion of seismic observations: J. of Geophys., v. 37, p. 225–235.
Musgrave, A. W., 1964, Wavefront charts and three-dimensional migrations: Geophysics, v. 26, p. 738–753.
Nakashima, K., 1977a, Computation of interval velocities from normal moveout velocities for multilayers with arbitrary curvatures in two dimensions: Report of the Tech. Res. Center, no. 4, Japan Petr. Dev. Corp., Tokyo (in Japanese).
——— 1977b, Errors in interval velocities obtained from stacking velocities for a multilayered structure: Report of the Tech. Res. Center, no. 4, Japan Petr. Dev. Corp. Tokyo (in Japanese).
Neidell, N. S., and Taner, M. T., 1971, Semblance and other coherency measures for multichannel data: Geophysics, v. 36, p. 482–497.
Newman, P., 1973, Divergence effects in a layered earth: Geophysics, v. 38, p. 481–488.
——— 1975, Amplitude and phase properties of a digital migration process: Paper presented at 37th Meeting of EAEG, Bergen, Norway.
O'Brien, P. N. S., and Lucas, A. L., 1971, Velocity dispersion of seismic waves: Geophys. Prosp., v. 19, p. 1–26.
Ortega, J. M., and Rheinboldt, W. C., 1970, Numerical iterative solution of non-linear equations in several variables: Computer Sci. Appl. Math. Series, New York, Academic Press.
O'Doherty, R. F., and Anstey, N. A., 1971, Reflections on amplitudes: Geophys. Prosp., v. 19, p. 430–458.

Officer, Ch. B., 1974, Introduction to theoretical geophysics: New York, Springer Verlag.
Paturet, D., 1977, Determination of high and low frequency static corrections: Paper presented at the 39th EAEG Meeting, Zagreb.
Pollet, A., 1974, Simple velocity modeling and continuous velocity section: Paper presented at the 44th Annual International SEG Meeting November 13, in Dallas.
Popov, M. M., and Pšenčik, I., 1976, Ray amplitudes in inhomogeneous media with curved interfaces: Traveaux Inst. Geophys. Acad. Tchechosl. Sci. No. 454, Geofysikalni sbornik, Academia, Praha, p. 111–129.
——— 1978, Computation of ray amplitudes in inhomogeneous media with curved interfaces: Studia Geoph. et Geod., v. 22, p. 248–258.
Postma, G. W., 1955, Wave propagation in a stratified medium: Geophysics, v. 20, p. 780–806.
Prescott, H. R., and Scanlan, 1971, The effects of weathering on derived velocities: Paper presented at the 41st International SEG Meeting November 10, in Houston.
Press, F., 1966, Seismic velocities, *in* Handbook of physical constants (rev. ed.): GSA memoir 97, p. 195–218.
Pšenčik, I., 1972, Kinematics of refracted and reflected waves in inhomogeneous media with non-planar interfaces: Studia Geoph. et Geod., v. 16, p. 26–52.
Rice, R. B., 1949, A discussion of steep-dip seismic computing methods, Part I: Geophysics, v. 14, p. 109–122.
——— 1950, A discussion of steep-dip seismic computing methods, Part II: Geophysics 15, p. 80–93.
Robinson, J. C., 1970a, An investigation of the relative accuracy of the most common normal moveout expression in velocity analyses: Geophys. Prosp., v. 18, p. 352–363.
——— 1970b, Statistically optimal stacking of seismic data: Geophysics, v. 35, p. 436–446.
Robinson, J. C., and Aldrich, C. A., 1972, A novel high-speed algorithm for summational type seismic velocity analyses: Geophys. Prosp., v. 20, p. 814–827.
Rockwell, D. W., 1971, Migration stack aids interpretation: Oil and Gas Journal, v. 69, April 19, 1971, p. 202–218.
Sághy, G., and Zelei, A., 1975, Advanced method for self-adaptive estimation of residual static corrections: Geophys. Prosp., v. 23, p. 259–271.
Sattlegger, J., 1964, Series for three-dimensional migration in reflection seismic interpretation: Geophys. Prosp., v. 12, p. 115–134.
——— 1965, A method of computing true interval velocities from expanding spread data in the case of arbitrary long spreads and arbitrarily dipping plane interfaces, Geophys. Prosp. 13, p. 306–318.
——— 1969, Three-dimensional seismic depth computation using space-sampled velocity logs, Geophysics, v. 34, p. 7–20.
Sattlegger, J., and Stiller, P. K., 1973, Section migration, before stack, after stack, or inbetween: Geophys. Prosp., v. 22, p. 297–314.
——— 1975, Migration velocity determination, Part I: Philosophy: Geophysics, v. 40, p. 1–5.
——— 1977, Downward continuation of time contour maps: Paper presented at the 39th EAEG Meeting, Zagreb.
Schmitt, A. 1966, Propagation of elastic waves in layered media: M.S. thesis, The University of Texas.
Schneider, J., 1979, Modeling techniques—Interpretational tools in seismic processing work: Presented at the 41st EAEG Meeting in Hamburg.
Schneider, W. A., 1971, Developments in seismic data processing and analysis (1968–1970): Geophysics, v. 36, p. 1043–1073.
Schneider, W. A., and Backus, M. M., 1968, Dynamic correlation analysis: Geophysics, v. 33, p. 105–126.
——— 1978, Integral formulation for migration in two and three dimensions: Geophysics, v. 43, p. 49–76.
Schultz, P., and Sherwood, J., 1978, Depth migration before stack: Paper presented at the 48th International SEG Meeting October 30, in San Francisco.
Shah, P. M., 1973a, Ray tracing in three dimensions: Geophysics, v. 38, p. 600–604.
——— 1973b, Use of wave front curvature to relate seismic data with subsurface parameters: Geophysics, v. 38, p. 812–825.
Shah, P. M., and Levin, F. K., 1973, Gross properties of time-distance curves: Geophysics, v. 38, p. 643–656.
——— 1974, Asymptotic ray method for seismic modeling: Paper presented at the Annual International SEG Meeting November 13, in Dallas.
Sheriff, R. E., 1975, Factors affecting seismic amplitudes: Geophys. Prosp., v. 23, p. 125–138.
Sherwood, J. W. C., and Poe, P. H. 1972, Continuous velocity estimation and seismic

wavelet processing: Geophysics, v. 37, p. 769–787.
Shugart, T. R., 1969, A critique of delta-T velocity determinations and a method for automating them: Geophysical Society of Houston.
Slotnick, M. M., 1936, On seismic computations, with applications, part I: Geophysics, v. 1, p. 9–22.
——— 1959, Lessons in seismic computing, A memorial to the author: SEG spec. publ., Tulsa.
Smith, S. G., 1977, A reflection profile modeling system: Geophys. J.R. Astr. Soc., v. 49, p. 723–737.
Sorrells, G. G., Crowley, J. B., and Veith, K. F., 1971, Methods for computing ray paths in complex geological structures: SSA Bull., v. 61, p. 27–53.
Späth, H., 1973, Spline Algorithmen zur Konstruktion glatter Kurven und Flächen: Oldenbourgh Verlag.
Stavroudis, O. N., 1972, The optics of rays, wave fronts, and caustics: New York, Academic Press.
——— 1976, Simpler derivation of the formulas for generalized ray tracing: J. Opt. Soc. Am., v. 66, p. 1330–1333.
Stolt, R. H., 1978, Migration by Fourier transform: Geophysics, v. 43, p. 23–48.
Stone, D. G., 1974, Velocity and bandwidth: Paper presented at 44th Annual International SEG Meeting, November 13, in Dallas.
Taner, M. T., and Koehler, F., 1969, Velocity spectra-digital computer derivation and applications of velocity functions: Geophysics, v. 34, p. 859–881.
Taner, M. T., Cook, E. E., and Neidell, N. S., 1970, Limitations of the reflection seismic method, lessons from computer simulations: Geophysics, v. 35, p. 551–573.
Taner, M. T., Koehler, F., and Alhilali, K. A., 1974, Estimation and correction of near-surface time anomalies: Geophysics, v. 39, p. 441–463.
Thornburg, H. R., 1930, Wave front diagrams in seismic interpretation: AAPG Bull., v. 14, p. 185–200.
Thomas, J. H., and Lucas, A. L., 1977, The effect of velocity anisotropy on stacking velocities and time to depth conversion: Paper presented at the 31st EAEG Meeting, Zagreb.
Timur, A., 1977, Temperature dependence of compressional and shear-wave velocities in rocks: Geophysics, v. 42, p. 950–956.
Toksöz, M. N., Cheng, Ch. H., and Timur, A., 1976, Velocities of seismic waves in porous rocks: Geophysics, v. 41, p. 621–645.
Trorey, A. W., 1970, A simple theory for seismic diffractions: Geophysics, v. 35, p. 762–784.
Tuchel, G., 1943, Seismische Messungen: Taschenbuch für Angewandte Geophysik, Leipzig.
Uhrig, L. F., and Van Melle, F. A., 1955, Velocity anisotropy in stratified media: Geophysics, v. 20, p. 774–779.
Ursin, B., 1977, Seismic velocity estimation: Geophys. Prosp., v. 25, p. 658–666.
Vlaar, N. J., 1968, Ray theory for an anisotropic inhomogeneous elastic medium: SSA Bull., v. 58, p. 2053–2072.
Walter, W. C., and Peterson, R. A., 1976, Seismic imaging atlas 1976: United Geophysical Corp. Publ., Pasadena.
White, J. E., 1965, Seismic waves, radiation, transmission, and attenuation: New York, McGraw-Hill Book Co., Inc.
White, R. E., 1977, The performance of optimum stacking filters in suppressing uncorrelated noise: Geophys. Prosp., v. 25, p. 165–178.
Wiggins, R. A., Larner, K. L., and Wisecup, R. D., 1976, Residual statics analysis as a general linear inverse problem: Geophysics, v. 41, p. 922–938.
Woods, John P., 1975, A seismic model using sound waves in air: Geophysics, v. 40, p. 593–607.
Wyllie, M. R. J., Gregory, A. R., and Gardner, L. W., 1956, Elastic wave velocities in heterogeneous and porous media: Geophysics, v. 21, p. 41–70.
Wyllie, M. R. J., Gregory, A. R., and Gardner, G. H. F., 1958, An experimental investigation of factors affecting elastic wave velocities in porous media: Geophysics, v. 23, p. 459–493.
Yacoub, N. K., Scott, J. H., and McKeown, F. A., 1970, Computer ray tracing through complex geological models for ground motion studies: Geophysics, v. 35, p. 586–602.
Zoeppitz, K., 1979, Erdbebenwellen VIII B: Über Reflexion und Durchgang seismischer Wellen durch Unstetigkeitsflächen: Gött. Nachr., v. 1, p. 66–84.

Index